Continuous Signals and Systems with MATLAB®

Electrical Engineering Textbook Series

Continuous Signals and Systems with MATLAB®, Third Edition

Volumes in the series:

Volume I – Continuous Signals and Systems with MATLAB®
Taan S. ElAli

Volume II – Discrete Signals and Systems with MATLAB®
Taan S. ElAli

Continuous Signals and Systems with MATLAB®

Third Edition

Authored by

Taan S. ElAli

Embry-Riddle Aeronautical University

CRC Press
Taylor & Francis Group
Boca Raton London New York

CRC Press is an imprint of the
Taylor & Francis Group, an **informa** business

First edition published 2001
by CRC Press
6000 Broken Sound Parkway NW, Suite 300, Boca Raton, FL 33487-2742
and by CRC Press
2 Park Square, Milton Park, Abingdon, Oxon, OX14 4RN
© 2021 Taan S. ElAli

CRC Press is an imprint of Taylor & Francis Group, LLC

ISBN: 978-0-367-53359-5 (hbk)
ISBN: 978-0-367-54299-3 (pbk)
ISBN: 978-1-003-08858-5 (ebk)

Visit the Routledge/Resources: www.routledge.com/9780367533595

Dedication

This book is dedicated first to the glory of the Almighty God. It is dedicated next to my beloved parents, my father, Saeed, and my mother, Shandokha—may God have mercy on their souls. It is dedicated then to my wife, Salam; my beloved children, Mohammad, Nusayba, Ali, and Zayd; my brothers, Mohammad and Khaled; and my sisters, Sabha, Khulda, Miriam, and Fatma. I ask the Almighty God to have mercy on us and to bring peace, harmony, and justice to all.

– Taan ElAli

Contents

Preface

All books on linear systems for undergraduates cover discrete and continuous systems material together in a single volume. Such books also include topics in discrete and continuous filter design, and discrete and continuous state-space representations. However, with this magnitude of coverage, the student typically gets a little of both discrete and continuous linear systems but not enough of either. Minimal coverage of discrete linear systems material is acceptable provided that there is ample coverage of continuous linear systems. On the other hand, minimal coverage of continuous linear systems does no justice to either of the two areas. Under the best of circumstances, a student needs a solid background in both these subjects. It is no wonder that these two areas are now being taught separately in so many institutions.

Continuous linear systems is a broad topic in itself and merits a single book devoted to that material. The objective of this book is to present the required material that an undergraduate student needs to master this subject matter and the use of MATLAB® (The MathWorks Inc.) in solving problems in this area.

This book offers broad, detailed, focused comprehensive coverage of continuous linear systems, based on basic mathematical principles. It presents many solved problems from various engineering disciplines using analytical tools as well as MATLAB. This book is intended primarily for undergraduate junior and senior electrical, mechanical, aeronautical, and aerospace engineering students. Practicing engineers will also find this book useful.

This book is ideal for use in a one-semester course in continuous linear systems where the instructor can easily cover all of the chapters. Each chapter presents numerous examples that illustrate each concept. Most of the worked-out examples are first solved analytically and then solved using MATLAB in a clear and understandable fashion.

This book concentrates on explaining the subject matter with easy-to-follow mathematical development and numerous solved examples. The book covers traditional topics and includes an extensive coverage of state-space representation and analysis. The reader does not need to be fluent in MATLAB because the examples are presented in a self-explanatory way.

Three chapters (7, 8, and 9) on block diagrams, the design of analog filters, and nonlinear systems, in addition to MATLAB m-files for the first 6 chapters, are provided online by the publisher at www.routledge.com/9780367533595.

To the Instructor: In one semester, all chapters can be covered in the order they are presented. For additional topics the instructor decides to cover, three chapters on block diagrams, the design of analog filters, and nonlinear systems are provided by the publisher at www.routledge.com/9780367533595. MATLAB m-files for the first six chapters will be provided at this link too.

To the Student: Familiarity with calculus, differential equations, and basic dynamics is desirable. If and where certain background material must be presented, that background material is presented right before the topic under consideration. This unique "just-in-time approach" helps the student stay focused on the topic.

This book presents three forms of numerical solutions using MATLAB. The first form allows you to type any command at the MATLAB prompt and then press the Enter key to get the results. The second form is the MATLAB script, a set of MATLAB commands you type and save in a file. You can run this file by typing the file name at the MATLAB prompt and then pressing the Enter key. The third form is the MATLAB function form, where you create the function and run it in the same way you create and run the script file. The only difference is that the name of the MATLAB function should be the same as the file name. Three chapters (7, 8, and 9) on block diagrams, the design of analog filters, and nonlinear systems, in addition to MATLAB m-files for the first 6 chapters, are provided online by the publisher at www.routledge.com/9780367533595.

To the Practicing Engineer: The practicing engineer will find topics in this book useful. In real life systems are nonlinear, and this book describes, step by step, the process of linearizing nonlinear systems. MATLAB, an invaluable tool for the practicing engineer, is used in solving most of the problems. Three chapters (7, 8, and 9) on block diagrams, the design of analog filters, and nonlinear systems, in addition to MATLAB m-files for the first 6 chapters, are provided online by the publisher at www.routledge.com/9780367533595.

MATLAB® is a registered trademark of The Math Works, Inc. For product information, please contact:

The Math Works, Inc.
3 Apple Hill Drive
Natick, MA 01760-2098
Tel: 508-647-7000
Fax: 508-647-7001
E-mail: info@mathworks.com
Web: http://www.mathworks.com

About the Author

Taan S. ElAli, PhD is a full professor of electrical engineering. He is currently the chair of the Engineering Program at ERAU. He has served as the chairman of the curriculum committee of the new College of Engineering at King Faisal University (KFU) from August 2008, when the committee was formed, to August 2009, when the final draft of the curriculum and its quality assurance plan for the college of engineering was approved by the University Council, the president of KFU, and the Ministry of Higher Education. From August 2009 to August 2010, Dr. ElAli was responsible for the implementation of the college of engineering curriculum quality assurance plan and the chairman of the electrical engineering department. Dr. ElAli has worked full-time at several academic institutions in the United States of America for more than 26 years in the areas of curriculum development, accreditation, teaching, research, etc. He received his BS degree in electrical engineering in 1987 from the Ohio State University, Columbus, Ohio, USA, his MS degree in systems engineering in 1989 from Wright State University, Dayton, Ohio, USA, and his M.S. in applied mathematical systems, and his PhD in electrical engineering, with a specialization in systems, controls, and signal processing from the University of Dayton in the years 1991 and 1993, respectively. Dr. ElAli has discovered a new approach to dynamic system identification and arrived at new mathematical formula in an automatically controlled dynamic system. Dr. ElAli is the author and coauthor of 13 books and textbooks in the area of electrical engineering and engineering education. He has contributed many papers and conference presentations in the area of dynamic systems and signal processing. In his efforts to revolutionize the engineering education, Dr. ElAli has contributed greatly to the Project-Based Curriculum (PBC) approach to engineering education. This approach has led to the restructuring of one of the major courses in the engineering curriculum and thus to the restructuring of the textbooks used for the course. From this restructuring, two textbooks have emerged: *Continuous Signals and Systems with MATLAB®* and *Discrete Systems and Digital Signal Processing with MATLAB®*. Universities like Rochester Institute of Technology, University of Texas at San Antonio, the University of Georgia, New Mexico State University, University of Massachusetts at Dartmouth, Texas A&M University at Kingsville, Western New England College, Lakehead University, University of New Haven, etc. have come on board to use these textbooks. He has taken the PBC approach to engineering education one step further and introduced student-focused learning through PBC. Dr. ElAli has contributed a chapter to the well-known *The Engineering Handbook* by CRC Press, another to the well-known *The Electrical Engineering Handbook*, by Taylor & Francis Group, and another two to *Advances in Systems*, and *Computing Sciences and Software Engineering*, by Springer, Germany, 2006 and 2007, respectively. Dr. ElAli is a Fulbright Scholar, and a senior member of the Institute of Electrical and Electronics Engineers (IEEE), a rare honor attained by fewer than 8% of IEEEs more than 400,000 members and is conferred only on those who have outstanding research achievements and who have performed great service to the scientific community.

Acknowledgment

We would like to thank the CRC Press team. Special thanks also go to Nora Konopka, who greatly encouraged us when we discussed this project with her initially. She has reaffirmed our belief that this is a much-needed book. Those thanks extend to all the people from CRC Press who worked on and helped in this project.

Finally, "nothing is perfect." Please forward any comments or concerns about this book to the publisher.

1 Signal Representation

1.1 EXAMPLES OF CONTINUOUS SIGNALS

We experience signals of various types almost on a continual basis in our daily lives. The blowing of the wind is an example of a continuous wave. We can plot the strength of the wind wave, the velocity of the wave, and the distance it travels as functions of time. When we speak, continuous signals are generated. These spoken word signals travel from one place to another so that other people can hear them. These are our familiar sound waves.

When a radar system detects a certain object in the sky, an electromagnetic signal is sent. This signal leaves the radar system and travels the distance in the air until it hits the target object, which then reflects back to the sending radar to be analyzed, where it is decided if the target is present or not. We understand that this electromagnetic signal, whether it is the one being sent or the one being received by the radar, is attenuated (its strength reduced) as it travels away from the radar station. Thus, the attenuation of this electromagnetic signal can be plotted as a function of time. If you vertically attach a certain mass to a spring from one end while the other end is fixed and then pull the mass, oscillations are created such that the spring's length increases and decreases until the oscillations finally stop. The oscillations produced are a signal that also dies out with increasing time. This signal, for example, can represent the length of the spring as a function of time.

Signals also can appear as electric waves. Examples are voltages and currents on long transmission lines. Voltage value is reduced as the impressed voltage travels on transmission lines from one city to another. Therefore, we can represent these voltages as signals as well and plot them in terms of time. When we discharge or charge a capacitor, the rate of charging and discharging depends on the time factor (among other factors). Charging and discharging the capacitor can be represented thus as voltage across the capacitor terminal as a function of time. These are a few examples of continuous signals that exist in nature and that can be modeled mathematically as signals that are functions of various parameters.

1.2 THE CONTINUOUS SIGNAL

A signal $v(t)$ is said to be continuous at time $t = t_1$, if and when $v\left(t_1^+\right) = v\left(t_1^-\right) = v(t_1)$. Where t_1^+ indicates time just after $t = t_1$ and t_1^- indicates time just before $t = t_1$. If, on the other hand, this is untrue, the signal $v(t)$ is referred to as a discontinuous signal. Figure 1.1 shows an example of a continuous signal.

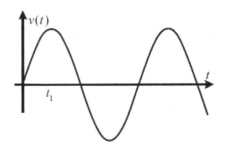

FIGURE 1.1   An example of a continuous signal.

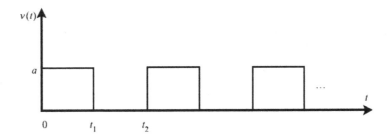

FIGURE 1.2   Signal for Example 1.1.

Example 1.1

Consider a signal of magnitude a as shown in Figure 1.2 and discuss its continuity.

SOLUTION

As you can see in Figure 1.2, $v(t_1^+) \neq v(t_1^-)$ since $v(t_1^+) = 0$ and $v(t_1^-) = a$. Also, $v(t_2^+) \neq v(t_2^-)$. Accordingly, in this example, the signal is piecewise continuous. It is continuous at all t except at $t = t_k$, where $k = 0, 1, 2, 3, \ldots$. It is customary to estimate a unique value for the signal at each of the discontinuities. For example, the value of $v(t)$ at $t = t_1$ is

$$v(t = t_1) = 1/2\left[v(t_1^+) + v(t_1^-)\right] = a/2.$$

1.3   PERIODIC AND NONPERIODIC SIGNALS

A signal $v(t)$ is periodic if $v(t) = v(t + kT)$, where k is an integer that has the values $\pm 1, \pm 2, \pm 3, \pm 4, \ldots$, and T is referred to as the fundamental period. In other words, if the signal $v(t)$ repeats itself every T seconds or units of time it is called periodic with period T. If $v_1(t)$ is periodic with period T_1 and if $v_2(t)$ is periodic with period T_2, what can we say about the period of $v(t) = v_1(t) + v_2(t)$? Consider $v(t)$ to be periodic having period T. Accordingly, we can say that $v(t) = v(t + T)$ as per the definition of periodicity just given to us.

Since $v_1(t)$ and $v_2(t)$ are each periodic with period T_1 and T_2, respectively, then

$$v_1(t) = v_1(t + kT_1),$$

and

$$v_2(t) = v_2(t + mT_2),$$

where k and m are integers. Since

$$v(t) = v_1(t) + v_2(t),$$

we have

$$v(t + T) = v_1(t + kT_1) + v_2(t + mT_2).$$

But

$$v_1(t + T) + v_2(t + T) = v_1(t + kT_1) + v_2(t + mT_2).$$

For this last equation to be true, we must have

$$T = kT_1$$

and

$$T = mT_2.$$

This implies

$$kT_1 = mT_2,$$

which says that

$$T_1 / T_2 = m / k$$

where m/k is a rational number. Therefore, the combination signal

$$v(t) = v_1(t) + v_2(t)$$

is a periodic signal if $T_1/T_2 = m/k$, where m/k is a rational number, and the period of $v(t)$ is

$$T = kT_1 = mT_2.$$

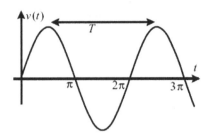

FIGURE 1.3   Sinusoidal signal.

1.4   GENERAL FORM OF SINUSOIDAL SIGNALS

Consider the sinusoidal signal shown in Figure 1.3. The magnitude of the signal $v(t)$ at $\pi/2$ and $5\pi/2$ is the same. The time between these values is called the period T. Every T units of time the signal repeats with the same magnitude. The period of this signal is 2π units of time. A general form of a sinusoidal signal is

$$v(t) = A\sin(\omega t + \Phi) \tag{1.1}$$

where
  A is the magnitude of the signal $v(t)$.
  ω is the angular frequency ($\omega = 2\pi f$).
  f is the frequency in hertz ($f = 1/T$).
  Φ is the phase angle in radians.

Example 1.2

Let $v_1(t) = \sin(t)$ and $v_2(t) = \cos(t)$. Consider the sum of the two signals $v(t) = v_1(t) + v_2(t)$. Is $v(t)$ periodic? If so, what is the period of $v(t)$?

SOLUTION

$v_1(t)$ has an angular frequency of 1 rad/sec and $v_2(t)$ has an angular frequency of 1 rad/sec, too. For $v_1(t)$ and with $w = 1$ we can write

$$1 = 2\pi/T_1 \Rightarrow T_1 = 2\pi.$$

Also, for $v_2(t)$ and with $w = 1$ we can write

$$1 = 2\pi/T_2 \Rightarrow T_2 = 2\pi.$$

Checking the ratio

$$T_1/T_2 = 2\pi/2\pi = 1$$

indicates that it is a rational number. Therefore, $v(t)$ is a periodic signal and has a period $T = kT_1 = mT_2$

$$T_1/T_2 = m/k = 1/1 \Rightarrow m = 1 \text{ and } k = 1.$$

Thus,

$$T = kT_1 = 2\pi$$

or

$$T = mT_2 = 2\pi.$$

Example 1.3

Let $v_1(t) = \sin(2\pi t)$ and $v_2(t) = \sin(t)$. Consider the sum $v(t) = v_1(t) + v_2(t)$. Is $v(t)$ periodic? If so, what is the period of $v(t)$?

SOLUTION

For $v_1(t)$ we have

$$T_1 = 1,$$

and for $v_2(t)$ we have

$$T_2 = 2\pi.$$

The ratio

$$T_1/T_2 = 1/2\pi$$

is not a rational number. Therefore, $v(t) = v_1(t) + v_2(t)$ is not a periodic signal.

1.5 ENERGY AND POWER SIGNALS

The signal energy over a time interval $2L$ is given by

$$E = \int_{-L}^{L} |v(t)|^2 \, dt. \qquad (1.2)$$

If the interval is infinite then

$$E = \int_{-\infty}^{\infty} |v(t)|^2 \, dt \qquad (1.3)$$

and in terms of limits, it is given as

$$E = \lim_{L \to \infty} \int_{-L}^{L} |v(t)|^2 \, dt. \qquad (1.4)$$

In Equation (1.4), $v(t)$ can be thought of as the voltage across a 1-Ohm resistor, and thus $(v(t))^2$ is the instantaneous power.

The average power is thus given by

$$P = \lim_{L \to \infty} (1/2L) \int_{-L}^{L} |v(t)|^2 \, dt. \qquad (1.5)$$

Let us consider the following two cases.

1. If in Equation (1.4) we find that $0 < E < \infty$, then $v(t)$ is referred to as the energy signal and the average power is zero.
2. If in Equation (1.5) we find that $0 < P < \infty$, then $v(t)$ is referred to as a power signal and it produces infinite energy.

For periodic signals, if we set $2L$ in Equation (1.5) such that $2L = mT$, where m is an integer then we have

$$P = \lim_{m \to \infty} (m/mT) \int_{0}^{T} |v(t)|^2 \, dt$$

which simplifies to

$$P = (1/T) \int_{0}^{T} |v(t)|^2 \, dt. \qquad (1.6)$$

Example 1.4

Let $v(t) = Ae^{-t}$ for $t \geq 0$. What kind of signal is given in Figure 1.4?

SOLUTION

Let us calculate the energy in the signal first as

$$E = \int_{0}^{\infty} |Ae^{-t}|^2 \, dt$$

$$E = \int_{0}^{\infty} A^2 e^{-2t} \, dt.$$

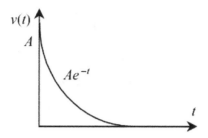

FIGURE 1.4 Signal for Example 1.4.

Carrying out the integration we get

$$E = \left(A^2/-2\right)e^{-2t}\Big|_0^\infty = \left(A^2/2\right).$$

Since $0 < E = A^2/2 < \infty$, $v(t)$ is an energy signal.

1.6 THE SHIFTING OPERATION

Consider the signal $v(t)$ in general. The representation $v(t - t_0)$ implies that $v(t)$ is shifted to the right t_0 units provided $t_0 > 0$. On the other hand, when $t_0 < 0$, $v(t - t_0)$ implies that $v(t)$ is shifted to the left t_0 units.

Example 1.5

Consider the signal shown in Figure 1.5. Determine mathematically both $v(t - 2)$ and $v(t + 2)$.

SOLUTION

Mathematically the signal in Figure 1.5 can be expressed as

$$v(t) = \begin{cases} t & t \geq 0 \\ 0 & t < 0. \end{cases}$$

FIGURE 1.5 Signal for Example 1.5.

To plot $v(t-2)$ we slide the graph of $v(t)$ two units to the right. Likewise, to plot $v(t+2)$ we slide the graph of $v(t)$ two units to the left. Mathematically

$$v(t-2) = \begin{cases} t-2 & t-2 \geq 0 \\ 0 & t-2 < 0 \end{cases}$$

and finally

$$v(t-2) = \begin{cases} t-2 & t \geq 2 \\ 0 & t < 2 \end{cases}.$$

Similarly,

$$v(t+2) = \begin{cases} t+2 & t+2 \geq 0 \\ 0 & t+2 < 0 \end{cases},$$

and finally

$$v(t+2) = \begin{cases} t+2 & t \geq -2 \\ 0 & t < -2 \end{cases}.$$

The graphical and analytical solutions will readily agree.

1.7 THE REFLECTION OPERATION

Given a signal $v(t)$, $v(-t)$ is referred to as its reflection. At times, it is also called an image of $v(t)$.

Example 1.6

For the signal shown in Figure 1.6, plot and find analytical expressions for $v(-t)$ and $v(1-t)$.

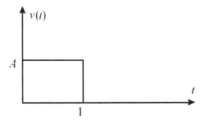

FIGURE 1.6   Signal for Example 1.6.

SOLUTION

Mathematically, the signal in Figure 1.6 can be expressed as

$$v(t) = \begin{cases} A & 0 \le t \le 1 \\ 0 & \text{otherwise} \end{cases}$$

and

$$v(-t) = \begin{cases} A & 0 \le -t \le 1 \\ 0 & \text{otherwise} \end{cases}.$$

By fixing the time range we get

$$v(-t) = \begin{cases} A & 0 \ge t \ge -1 \\ 0 & \text{otherwise} \end{cases}$$

or

$$v(-t) = \begin{cases} A & -1 \le t \le 0 \\ 0 & \text{otherwise} \end{cases}.$$

Likewise,

$$v(1-t) = \begin{cases} A & 0 \le 1-t \le 1 \\ 0 & \text{otherwise} \end{cases}.$$

By fixing the time interval we get

$$v(1-t) = \begin{cases} A & -1 \le -t \le 0 \\ 0 & \text{otherwise} \end{cases},$$

or

$$v(1-t) = \begin{cases} A & 1 \ge t \ge 0 \\ 0 & \text{otherwise} \end{cases},$$

and finally

$$v(1-t) = \begin{cases} A & 0 \le t \le 1 \\ 0 & \text{otherwise} \end{cases}.$$

The functions $v(-t)$ and $v(1 - t)$ are shown plotted in Figures 1.7 and 1.8, respectively.

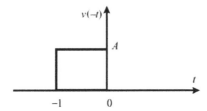

FIGURE 1.7   Signal for Example 1.6.

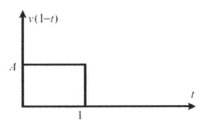

FIGURE 1.8   Signal for Example 1.6.

1.8   EVEN AND ODD FUNCTIONS

Consider the signal $v(t)$. It is said to be an even signal when $v(t) = v(-t)$. When $v(-t) = -v(t)$, then $v(t)$ is referred to as an odd signal. Any signal $v(t)$ can be represented as the sum of even and odd signals as

$$v(t) = v_{\text{even}}(t) + v_{\text{odd}}(t),$$

where

$$v_{\text{even}}(t) = 1/2[v(t) + v(-t)]$$

and

$$v_{\text{odd}}(t) = 1/2[v(t) - v(-t)].$$

Example 1.7

Consider the signals given in Figures 1.9 and 1.10. Are these signals odd, even, or neither?

SOLUTION

The signal shown in Figure 1.9 is odd since $v(t_1) = -v(-t_1)$. On the other hand, the signal shown in Figure 1.10 is even since $v(t_1) = v(-t_1)$.

FIGURE 1.9 Signal for Example 1.7.

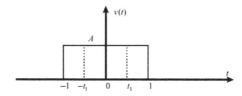

FIGURE 1.10 Signal for Example 1.7.

Example 1.8

Express the signal shown in Figure 1.11 as the sum of odd and even signals.

SOLUTION

As you can see, $v(t)$ is neither even nor odd. It can be written as the sum of an even and an odd signal, where

$$v_{even}(t) = 1/2[v(t) + v(-t)]$$

as shown in Figure 1.12, and

$$v_{odd}(t) = 1/2[v(t) - v(-t)],$$

and is shown plotted in Figure 1.13. The sum of these two signals is seen to resemble the signal given.

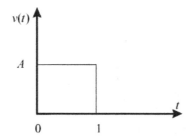

FIGURE 1.11 Signal for Example 1.8.

FIGURE 1.12   Signal for Example 1.8.

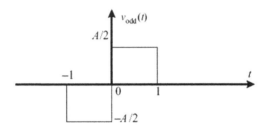

FIGURE 1.13   Signal for Example 1.8.

1.9   TIME SCALING

Given the signal $v(t)$, $v(nt)$ is a compressed signal of $v(t)$ if and when $|n| > 1$. On the other hand, $v(nt)$ becomes an expanded signal of $v(t)$ if and when $|n| < 1$.

Example 1.9

Given $v(t)$ shown in Figure 1.14a, find $v(t/2)$ and $v(2t)$.

SOLUTION

The solution is presented graphically in Figure 1.14b.

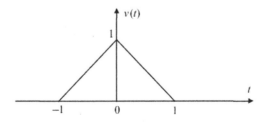

FIGURE 1.14a   Signal for Example 1.9.

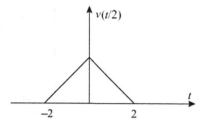

FIGURE 1.14b   Signal for Example 1.9.

Example 1.10

Consider the signal shown in Figure 1.15. Plot and analytically find $v(2t - 1)$.

SOLUTION

Mathematically, we can represent the signal shown in Figure 1.15 using the equation

$$v(t) = \begin{cases} t & 0 \le t \le 1 \\ 1 & 1 \le t \le 2 \\ 0 & t > 2 \end{cases}.$$

The signal shifted by 1 and scaled by 2 is

$$v(2t - t) = \begin{cases} 2t - 1 & 0 \le 2t - 1 \le 1 \\ 1 & 1 \le 2t - 1 \le 2 \\ 0 & 2t - 1 > 2 \end{cases}.$$

By fixing the time scale, we get

$$v(2t - t) = \begin{cases} 2t - 1 & 1 \le 2t \le 2 \\ 1 & 2 \le 2t \le 3 \\ 0 & 2t > 3 \end{cases},$$

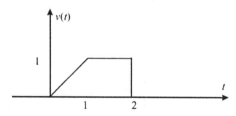

FIGURE 1.15   Signal for Example 1.10.

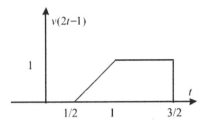

FIGURE 1.16   Signal for Example 1.10.

or finally

$$v(2t-1) = \begin{cases} 2t-1 & 1/2 \le t \le 1 \\ 1 & 1 \le t \le 3/2 \\ 0 & t > 3/2 \end{cases}.$$

The signal $v(2t-1)$ is thus obtained from $v(t)$ by compressing its slant segment by 1/2 and shifting the constant segment by $t = 1/2$. The resulting graph of $v(2t-1)$ is shown plotted in Figure 1.16.

1.10   THE UNIT STEP SIGNAL

It is common to close and open an electrical switch to cause a sudden change in the signal of an electrical circuit. To effect changes in a fluid-based system you may actuate a valve. You may also release a spring, which was compressed in a mechanical system. In these and similar situations, the signal value changes quickly as compared to the response of the system. This is what we commonly refer to as the step signal. Mathematically, a step signal such as that shown in Figure 1.17 is represented as

$$Au(t) = \begin{cases} A & t > 0 \\ \dfrac{A}{2} & t = 0 \\ 0 & t < 0 \end{cases}$$

where A is the magnitude of the step signal. The signal is discontinuous at $t = 0$. $u(t)$ is like a switch that is on for $t > 0$.

FIGURE 1.17   The step signal.

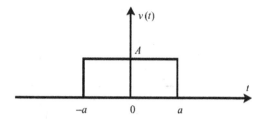

FIGURE 1.18 Signal for Example 1.11.

Example 1.11

Express the rectangular pulse signal shown in Figure 1.18 as a sum of unit step signals.

SOLUTION

Let us look at the plots of $Au(t + a)$ and $Au(t - a)$ shown in Figures 1.19 and 1.20. By subtracting the signal of Figure 1.20 from that of Figure 1.19, we can write

$$Au(t + a) - Au(t - a) = A\left[u(t + a) - u(t - a)\right].$$

The graph of the difference shown in Figure 1.21 is identical to the rectangular pulse given in Figure 1.18.

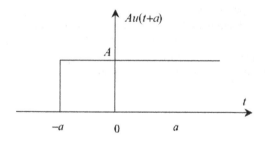

FIGURE 1.19 Signal for Example 1.11.

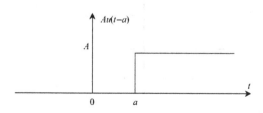

FIGURE 1.20 Signal for Example 1.11.

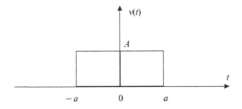

FIGURE 1.21   Signal for Example 1.11.

1.11   THE SIGNUM SIGNAL

The signum signal is sketched in Figure 1.22. Mathematically, it can be expressed as

$$sgn(t) = \begin{cases} 1 & t > 0 \\ 0 & t = 0 \\ -1 & t < 0 \end{cases}.$$

Is $sgn(t) = -1 + 2u(t)$? To check this, let us look at the following cases. If $t > 0$, $u(t) = 1$. In this case $-1 + 2(1) = 1$. When $t = 0$, $u(t) = 1/2$. Herein, $-1 + 2(1/2) = 0$. Finally, when $t < 0$, $u(t) = 0$. In this case $-1 + 2(0) = -1$. These cases do conform to the graph. Therefore, indeed, $sgn(t) = -1 + 2u(t)$.

1.12   THE RAMP SIGNAL

Ramp signals are also commonplace in many physical systems. A ramp signal can exist in a tracking-type situation, for example, when a target is moving at a constant velocity. A tracking antenna may need a ramp signal in order to track a plane traveling at a constant speed. In this case, the ramp signal increases linearly as a function of time as shown in the plot in Figure 1.23.

Mathematically, the ramp signal can be written as

$$r(t) = \begin{cases} t & t \geq 0 \\ 0 & t < 0 \end{cases}.$$

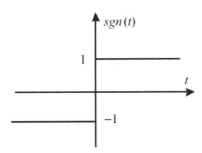

FIGURE 1.22   The signum signal.

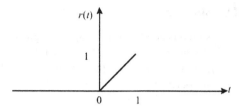

FIGURE 1.23 The ramp signal.

Notice that

$$\int_{-\infty}^{t} u(\tau)d\tau = \int_{0}^{t} 1 d\tau = \begin{cases} t & t \geq 0 \\ 0 & t < 0 \end{cases},$$

which is the same as the ramp signal described earlier. Accordingly, we can also write

$$\int_{-\infty}^{t} u(\tau)d\tau = r(t).$$

1.13   THE SAMPLING SIGNAL

The sampling signal shown in Figure 1.24 can be expressed mathematically as

$$S_a(x) = \frac{\sin(x)}{x}.$$

The *sinc* signal is

$$sinc(x) = \frac{\sin(\pi x)}{\pi x}.$$

The *sinc* signal is the sampling signal evaluated at $x = \pi x$ and written as

$$sinc(x) = S_a(\pi x).$$

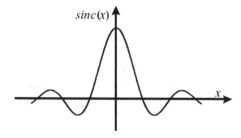

FIGURE 1.24 The sampling signal.

1.14 THE IMPULSE SIGNAL

The derivative of the step signal is defined as the impulse signal. Since the step signal is discontinuous at $t = 0$, the derivative of the step is of infinite magnitude at $t = 0$. This is shown in Figure 1.25, where A is the strength of the signal. The impulse signal $\delta(t)$ has the following properties:

1. $\delta(t) = 0$ for $t \neq 0$
2. $\delta(0) \to \infty$
3. $\displaystyle\int_{-\infty}^{+\infty} \delta(t)\,dt = 1$
4. $\delta(t)$ is even; $\delta(t) = \delta(-t)$

Evaluating an integral where the integrand is multiplied by an impulse function is relatively easy if we use the sifting property of the impulse signal as

$$\int_{t_1}^{t_2} v(t)\delta(t - t_0)\,dt = \begin{cases} v(t_0) & t_1 < t_0 < t_2 \\ 0 & \text{otherwise} \\ (1/2)v(t_0) & t_0 = t_1,\ t_0 = t_2 \end{cases}.$$

The scaling property of impulse signals is given by

$$\delta(at + b) = \frac{1}{|a|}\delta\left(t + \frac{b}{a}\right).$$

It can be used to evaluate an integral of $v(t)\delta(at + b)$ as

$$\int_{t_1}^{t_2} v(t)\delta(at + b)\,dt = \int_{t_1}^{t_2} \frac{1}{|a|}v(t)\delta\left(t + \frac{b}{a}\right)dt = \frac{1}{|a|}v\left(\frac{-b}{a}\right) \quad t_1 < \frac{-b}{a} < t_2$$

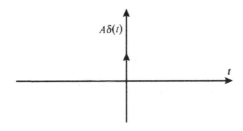

FIGURE 1.25 The impulse signal.

Example 1.12

Use the sifting property to evaluate the following integrals:

1. $\displaystyle\int_{0}^{7} (t+1)\delta(t-1)\,dt$

2. $\displaystyle\int_{0}^{\infty} e^{-t}\delta(t-10)\,dt$

3. $\displaystyle\int_{-\infty}^{t} \delta(\tau)\,dt$

SOLUTION

1. Since $0 < 1 < 7$ is true and because the impulse is applied at $t = 1$, the integral evaluates to

$$\int_{0}^{7} (t+1)\delta(t-1)\,dt = 2.$$

2. Since $0 < 10 < \infty$ is true and because the impulse is applied at $t = 10$, the integral reduces to

$$\int_{0}^{\infty} e^{-t}\delta(t-10)\,dt = e^{-10}.$$

3. For $\displaystyle\int_{-\infty}^{t} \delta(\tau)\,dt,$ we look at the following cases:

   a. if $t < 0$, the point, $\tau = 0$, is not between $-\infty$ and t.
   b. if $t > 0$, the point, $\tau = 0$, is between $-\infty$ and t.

Therefore,

$$\int_{-\infty}^{t} \delta(\tau)\,d\tau = \begin{cases} 1 & t > 0 \\ 0 & t < 0 \end{cases}.$$

What you see in the above equation is in fact the definition of the unit step signal, $u(t)$. This is to say that the integral of the impulse signal is the unit step signal.

1.15   SOME INSIGHTS: SIGNALS IN THE REAL WORLD

The signals that we have introduced in this chapter were all represented in mathematical form and plotted on graphs. This is how we represent signals for the purpose of analysis, synthesis, and design. For better understanding of these signals, we will provide real-life situations and examine the relation between these mathematical abstractions of signals and what they may represent in real life.

1.15.1   THE STEP SIGNAL

The mathematical abstraction is repeated in Figure 1.26 where

$$Au(t) = \begin{cases} A & t > 0 \\ 0 & t < 0 \end{cases}.$$

In real-life situations, this signal can be viewed as a constant force of magnitude A newtons applied at time equals zero seconds to a certain object for a long time. In another situation, $Au(t)$ can be an applied voltage of constant magnitude to a load resistor R at the time $t = 0$.

1.15.2   THE IMPULSE SIGNAL

The mathematical abstraction is repeated in Figure 1.27.

Again, in real life, this signal can represent a situation where a person hits an object with a hammer with a force of A newtons for a very short period of time (picoseconds). We sometimes refer to this kind of signal as a shock.

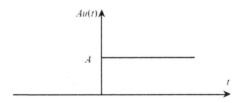

FIGURE 1.26   The step signal.

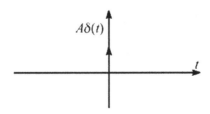

FIGURE 1.27   The impulse signal.

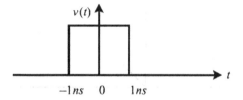

FIGURE 1.28 Modeling an impulse.

In another real-life situation, the impulse signal can be as simple as closing and opening an electrical switch in a very short time. Another situation where a spring-carrying mass is hit upward can be seen as an impulsive force. A sudden oil spill like the one that occurred in the Gulf War can represent a sudden flow of oil, which can also be seen as an impulse. You may realize that it is impossible to generate a pure impulse signal for zero duration and infinite magnitude. To create an approximation of an impulse, we can generate a pulse signal where the duration of the signal is very short compared to the response of the system. Such a pulse is often approximated as shown in the plot of Figure 1.28.

1.15.3 THE SINUSOIDAL SIGNAL

The mathematical abstraction of a sinusoidal signal can be represented again as shown in Figure 1.29 and mathematically we can write

$$v(t) = A\sin(\omega t)$$

where A is the maximum magnitude of the signal $v(t)$, ω is the radian frequency ($\omega = 2\pi f$), and f is the frequency in hertz ($f = 1/T$). This signal can be thought of as a situation where a person is shaking an object with a regular motion, where A is the magnitude of the force applied in newtons.

This is like pushing and pulling an object continuously with a period of T seconds. Thus, a push and pull forms a complete period of shaking. The distance the object covers during this shaking represents a sinusoidal signal. In the case of electrical signals, an AC voltage source is a sinusoidal signal.

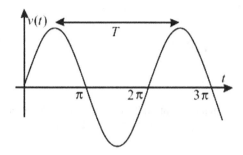

FIGURE 1.29 The sinusoidal signal.

FIGURE 1.30   The ramp signal.

1.15.4 THE RAMP SIGNAL

The ramp signal was represented earlier as a mathematical model and is repeated here, as shown in Figure 1.30, where

$$r(t) = \begin{cases} t & t \geq 0 \\ 0 & t < 0 \end{cases}.$$

In real-life situations, this signal can be viewed as a signal that is increasing linearly with time. An example is where a person starts applying a force to an object at time $t = 0$ and keeps pressing the object with increasing force for a long time. The rate of the increase in the force applied is constant.

Consider another situation where a radar system, an anti-aircraft gun, and an incoming jet interact. The radar antenna can provide an angular position input. Notion of the jet motion forces this angle to change uniformly with time. This will force a ramp input signal to the anti-aircraft gun since it will have to track the jet.

1.15.5 OTHER SIGNALS

A train of impulses can be thought of as hitting an object with a hammer continuously and uniformly. In terms of electricity, you may be continuously closing and opening a switch. A rectangular pulse is like applying a constant force to an object at a certain time and then removing that force. It is also like applying a constant voltage at a certain time and then closing the switch of the voltage source. Other signals are the random signals, where the magnitude changes randomly as time progresses. These signals can be thought of as shaking an object randomly with variable force. Gusting wind is another example of random signals.

1.16   END-OF-CHAPTER EXAMPLES

EOCE 1.1

A triangular voltage signal of duration 2 ms is applied as an input to an RC circuit. The input signal is given by $x(t)$ as described in Figure 1.31.

   1. Sketch $x(-t)$.
   2. Sketch $x(-3 - t)$.

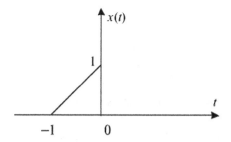

FIGURE 1.31 Signal for EOCE 1.1.

3. Sketch $x(t/3)$.
4. Find the even and the odd parts of $x(t)$ and sketch them.

SOLUTION

1. Mathematically, the signal $x(t)$ can be written as

$$x(t) = \begin{cases} t+1 & -1 < t < 0 \\ 0 & \text{elsewhere} \end{cases}.$$

The reflection of $x(t)$ is $x(-t)$ and is written as

$$x(-t) = \begin{cases} -t+1 & -1 < -< 0 \\ 0 & \text{elsewhere} \end{cases} = \begin{cases} -t+1 & 1 > t > 0 \\ 0 & \text{elsewhere} \end{cases},$$

or

$$x(-t) = \begin{cases} -t+1 & 0 < t < 1 \\ 0 & \text{elsewhere} \end{cases}.$$

$x(-t)$ is sketched in Figure 1.32.

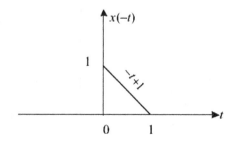

FIGURE 1.32 Signal for EOCE 1.1.

2. The signal

$$x(t) = \begin{cases} t+1 & -1 < t < 0 \\ 0 & \text{elsewhere} \end{cases}$$

reflected and translated three units is

$$x(-3-t) = \begin{cases} -3-t+1 & -1 < -3-t < 0 \\ 0 & \text{elsewhere} \end{cases}.$$

Fixing the time scale we get

$$x(-3-t) = \begin{cases} -2-t & -1+3 < -t < 0+3 \\ 0 & \text{elsewhere} \end{cases}$$

or

$$x(-3-t) = \begin{cases} -2-t & 2 < -t < 3 \\ 0 & \text{elsewhere} \end{cases}$$

$$x(-3-t) = \begin{cases} -2-t & 2 > -t > 3 \\ 0 & \text{elsewhere} \end{cases},$$

and finally

$$x(-3-t) = \begin{cases} -2-t & -3 < t < -2 \\ 0 & \text{elsewhere} \end{cases}.$$

This is a reflection and also a shifting operation involving $x(t)$. The sketch of $x(-3 -t)$ is shown in Figure 1.33.

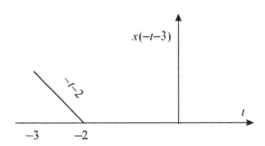

FIGURE 1.33 Signal for EOCE 1.1.

FIGURE 1.34 Signal for EOCE 1.1.

3. Rewriting $x(t)$ again we have

$$x(t) = \begin{cases} t+1 & -1 < t < 0 \\ 0 & \text{elsewhere} \end{cases}.$$

Scaling with a factor of 3 gives

$$x(t/3) = \begin{cases} t/3+1 & -1 < t/3 < 0 \\ 0 & \text{elsewhere} \end{cases},$$

and finally

$$x(t/3) = \begin{cases} t/3+1 & -3 < t < 0 \\ 0 & \text{elsewhere} \end{cases}.$$

The plot of $x(t/3)$ is shown in Figure 1.34.

4. Any signal $x(t)$ can be represented as the sum of even and odd signals as

$$x(t) = x_{even}(t) + x_{odd}(t),$$

Where

$$x_{even}(t) = 1/2\left[x(t) + x(-t)\right]$$

$$x_{even}(t) = \begin{cases} t/2+1/2 & -1 < t < 0 \\ 0 & \text{elsewhere} \end{cases} + \begin{cases} -t/2+1/2 & 0 < t < 1 \\ 0 & \text{elsewhere} \end{cases},$$

and the plot is given in Figure 1.35.

$$x_{odd}(t) = 1/2\left[x(t) - x(-t)\right]$$

$$x_{odd}(t) = \begin{cases} t/2+1/2 & -1 < t < 0 \\ 0 & \text{elsewhere} \end{cases} - \begin{cases} -t/2+1/2 & 0 < t < 1 \\ 0 & \text{elsewhere} \end{cases},$$

and the plot is given in Figure 1.36.

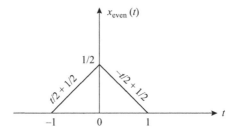

FIGURE 1.35   Signal for EOCE 1.1.

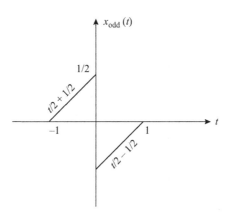

FIGURE 1.36   Signal for EOCE 1.1.

The sum of the odd and the even signals is obtained, as shown in Figure 1.37. Using MATLAB, we can simulate EOCE 1.1 using the m-file EOCE1_1.

and the plots are given in Figure 1.38.

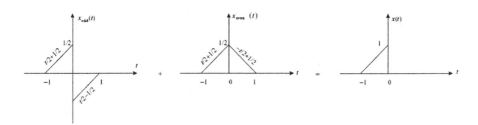

FIGURE 1.37   Signal for EOCE 1.1.

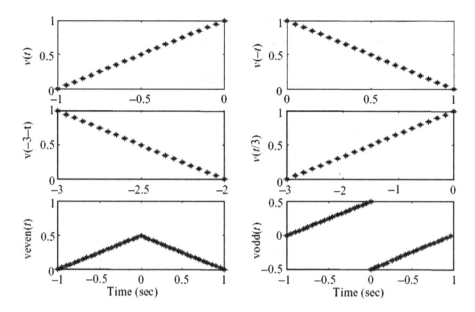

FIGURE 1.38 Signals for EOCE 1.1 as plotted by MATLAB.

EOCE 1.2

Consider the two signals as given in Figures 1.39 and 1.40. Are these signals power or energy signals?

FIGURE 1.39 Signal for EOCE 1.2.

FIGURE 1.40 Signal for EOCE 1.2.

SOLUTION

Mathematically, the first signal can be represented as

$$x_1(t) = \begin{cases} 10 & -5 < t < 5 \\ 0 & \text{otherwise} \end{cases},$$

and the second signal as

$$x_2(t) = \begin{cases} 2t & 0 < t < 5 \\ 0 & \text{otherwise} \end{cases}.$$

Energy for the first signal is calculated as

$$E_1 = \int_{-5}^{5} 100 \, dt = 100(5+5) = 1000,$$

and energy for the second signal is calculated as

$$E_2 = \int_{0}^{5} 4t^2 \, dt = \frac{4}{3}(125 - 0) = \frac{500}{3}.$$

Since E_1 and E_2 have finite energy, these signals are energy signals.

EOCE 1.3

Consider the following signals.

$$x_1(t) = 10\cos(2t)$$

$$x_2(t) = \cos(2\pi t)$$

$$x_3(t) = (1/2)\cos((2/3)t + 10)$$

$$x_4(t) = 10\cos(2t) + \sin(3\pi t)$$

1. What are the periods of the first three signals given?
2. Is the sum of the first two signals periodic?
3. Is the sum of the last two signals periodic?
4. Is the sum of the first and the last signals periodic?
5. Plot the signals in all parts.

SOLUTION

1. For the first signal: $w_1 = 2$ with $T1 = \pi$ sec.
   For the second signal: $w_2 = 2\pi$. Therefore, $T_2 = 1$ sec.
   For the third signal: $w_3 = 2/3$. Therefore, $T_3 = 3\pi$ sec.

2. $x_1(t) + x_2(t)$ is a periodic signal if $T_1/T_2 = m/k$, where m/k is a rational number and T_1 and T_2 are the periods of $x_1(t)$ and $x_2(t)$, respectively. In our case, $T_1 = \pi$ and $T_2 = 1$. $T_1/T_2 = \pi$, which is not a rational number, and hence $x_1(t) + x_2(t)$ is not periodic.
3. The period for the third signal is $T_3 = 3\pi$. The last signal has two parts (signals). The first part has a period of π sec and the second part has a period of 2/3 sec. Again, $\pi/(2/3) = (3\pi)/2$ is not a rational number. Therefore, signal four is not periodic, and so the sum of the last two signals is not periodic.
4. Since the last signal is not periodic, the sum of the first and the last is not periodic.

We will use MATLAB to plot the signals. The MATLAB script EOCE1_31 is used.

The plots are shown in Figure 1.41.

As you can see, the last signal is not periodic. We can write another script, EOCE1_32, to plot the other signals.

The plots are shown in Figure 1.42.

You also can see that the signals in Figure 1.42 are not periodic.

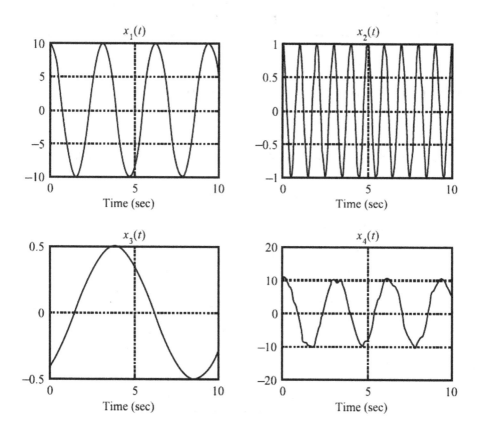

FIGURE 1.41   Signals for EOCE 1.3 as plotted by MATLAB.

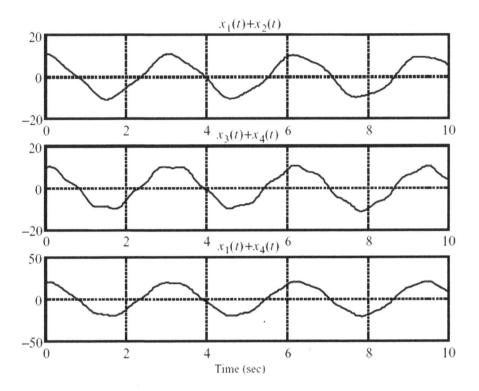

FIGURE 1.42   Signals for EOCE 1.3 as plotted by MATLAB.

EOCE 1.4

Consider the single-phase circuit shown in Figure 1.43. Let the input voltage be given as

$$v(t) = V_m \cos(wt + \theta_v).$$

and the current be given as

$$i(t) = I_m \cos(wt + \theta_i).$$

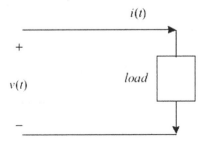

FIGURE 1.43   Circuit for EOCE 1.4.

For the purpose of calculations and plot, assume values for the variables used as

$$v(t) = 100\cos wt$$

$$i(t) = 80\cos(wt - 60)$$

$$p(t) = 800wt\cos(wt - 60).$$

Plot the curves for $i(t)$, $v(t)$, and $p(t)$, and show that $p(t)$ has a frequency twice the frequency of the input $v(t)$.

SOLUTION

We use the MATLAB script EOCE1_4 to produce the plots below.
The plots are given in Figure 1.44.
Notice that the frequency of the power $p(t)$ is twice the frequency of the voltage $v(t)$. The instantaneous power delivered to the load is calculated as

$$p(t) = V_m\cos(wt + \theta_v)I_m\cos(wt + \theta_i)$$

$$p(t) = \frac{1}{2}V_mI_m\left[\cos(\theta_v - \theta_i) + \cos(2wt + \theta_v + \theta_i)\right].$$

You also can see mathematically that the frequency of the instantaneous power $p(t)$ is twice the frequency of the source voltage $v(t)$ in the term $\cos(2wt + \theta_v + \theta_i)$.

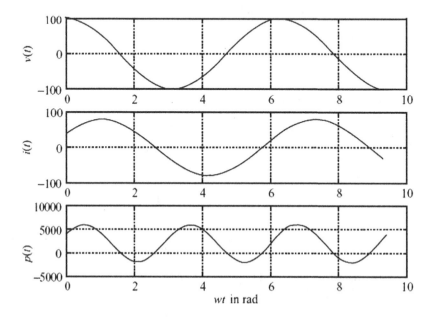

FIGURE 1.44 Signals for EOCE 1.4 as plotted by MATLAB.

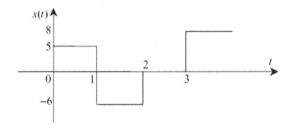

FIGURE 1.45   Signal for EOCE 1.5.

EOCE 1.5

Consider the signal shown in Figure 1.45.

    1. Write the signal as a sum of unit step signals.
    2. Knowing that the derivative of a unit step $u(t)$ is the impulse $\delta(t)$, as shown
       in Figure 1.46, what is the derivative of the given signal?

SOLUTION

1. The given signal can be written mathematically as

$$x(t) = 5u(t) - 11u(t-1) + 6u(t-2) + 8u(t-3).$$

2. The derivative of the signal is

$$\frac{d}{dt}x(t) = 5\delta(t) - 11\delta(t-1) + 6\delta(t-2) + 8\delta(t-3).$$

and is shown in Figure 1.47.

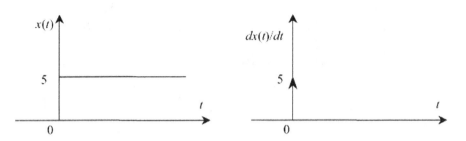

FIGURE 1.46   Signals for EOCE 1.5.

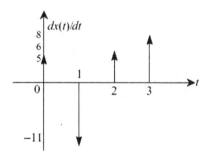

FIGURE 1.47 Signal for EOCE 1.5.

EOCE 1.6

Consider the signal shown in Figure 1.48.

1. Write the signal as a sum of unit step signals.
2. Knowing that the derivative of a unit step $u(t)$ is the impulse $\delta(t)$, what is the derivative of $x(t)$?

SOLUTION

1. Mathematically, $x(t)$ can be written as

$$x(t) = 5t\big[u(t) - u(t-1)\big] - 6\big[u(t-1) - u(t-2)\big]$$
$$+ (5t - 10)\big[u(t-2) - u(t-3)\big].$$

2. The derivative of $x(t)$ can be written as

$$\frac{d}{dt}x(t) = \big[5u(t) + 5t\delta(t) - 5u(t-1) - 5t\delta(t-1)\big]$$
$$+ \big[-6\delta(t-1) + 6\delta(t-2)\big]$$
$$+ \big[5u(t-2) + 5t\delta(t-2) - 5u(t-3)\big]$$
$$- \big[5t\delta(t-3)\big(-10\delta(t-2)\big) + 10\delta(t-3)\big]$$

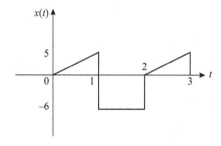

FIGURE 1.48 Signal for EOCE 1.6.

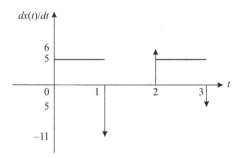

FIGURE 1.49   Signal for EOCE 1.6.

or

$$\frac{d}{dx}x(t) = 5u(t) - 5u(t-1) - 5\delta(t-1)$$

$$- 6\delta(t-1) + \big(6\delta(t-2) + 5u(t-2)\big)$$

$$+ 10\delta(t-2) - 5u(t-3) - 15\delta(t-3)$$

$$- 10\delta(t-2) + 10\delta(t-3).$$

Finally

$$\frac{d}{dx}x(t) = 5u(t) - 5u(t-1) - 11\delta(t-1) + 6\delta(t-2)$$

$$+ 5u(t-2) - 5u(t-3) - 5\delta(t-3)$$

and is shown in Figure 1.49.

EOCE 1.7

Usually, evaluating integrals is somewhat difficult in many cases. When the integrand is a multiplication of a function $x(t)$ and an impulse function $\delta(t)$, the integral can be evaluated easily using the following formula.

$$\int_{t_1}^{t_2} x(t)\delta(t-t_0)\,dt = \begin{cases} x(t_0) & t_1 < t_0 < t_2 \\ 0 & \text{otherwise} \\ (1/2)x(t_0) & t_0 = t_1,\ t_0 = t_2. \end{cases}$$

Find the following integrals.

1. $\displaystyle\int_0^{10} e^{-t^2}\delta(t-6)\,dt$

2. $\int\limits_{0}^{10} e^{-t^2} \cos(wt)\delta(t+6)\,dt$

3. $\int\limits_{0}^{10} e^{-t^2} x(t)\delta(t-6)\,dt$

SOLUTION

1. The impulse is applied at $t = 6$ units of time, which is within the limits of the integral.
2. The impulse is applied at $t = -6$ units of time, which is outside the limits of the integral.
3. The impulse is applied at $t = 6$ units of time, which is within the limits of the integral.

Therefore,

$$\int\limits_{0}^{10} e^{-t^2}\delta(t-6)\,dt = e^{-6^2} = e^{-36}$$

$$\int\limits_{0}^{10} e^{-t^2}\cos(wt)\delta(t+6)\,dt = 0$$

$$\int\limits_{0}^{10} e^{-t^2} x(t)\delta(t-6)\,dt = e^{-36}x(6).$$

EOCE 1.8

Use MATLAB to simulate the following signals.

1. The step signal, $u(t)$
2. The impulse signal, $\delta(t)$
3. The ramp signal, $r(t)$
4. A general sinusoidal signal

SOLUTION

1. The MATLAB script EOCE1_81 is used to simulate a unit step function with amplitude A that starts at time = 0.
   The plot for the unit step is shown in Figure 1.50.
2. The MATLAB script EOCE1_82 is used to simulate the impulse signal.
   The plot is shown in Figure 1.51.
3. The MATLAB script EOCE1_83 is used to simulate the ramp signal.
   The plot is shown in Figure 1.52.
4. The MATLAB script EOCE1_84 is used to simulate the general sinusoidal signal.
   The plot is shown in Figure 1.53.

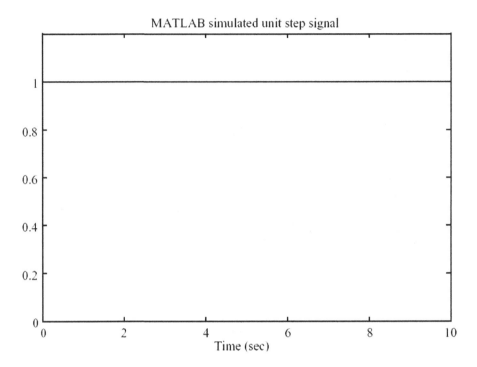

FIGURE 1.50   Signal for EOCE 1.8 as plotted by MATLAB.

FIGURE 1.51   Signal for EOCE 1.8 as plotted by MATLAB.

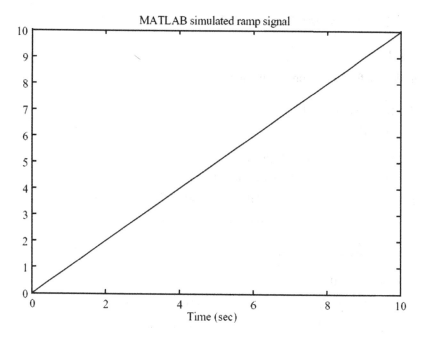

FIGURE 1.52   Signal for EOCE 1.8 as plotted by MATLAB.

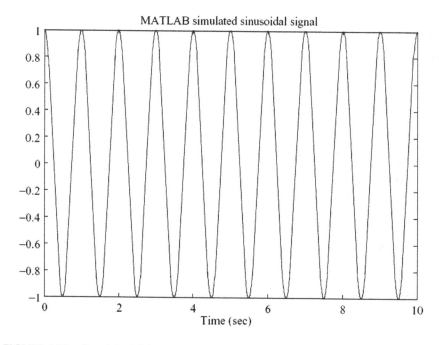

FIGURE 1.53   Signal for EOCE 1.8 as plotted by MATLAB.

1.17   END-OF-CHAPTER PROBLEMS

EOCP 1.1

Consider the following continuous signals.

1. $v(t) = 10t\cos(2t+1)$
2. $v(t) = 10\cos(2t+1) + 10$
3. $v(t) = \cos(5t+1) + 10 + \sin(t)e^{-t}$
4. $v(t) = 100\cos(2t+100) - 10 + \sin(3t-3)$
5. $v(t) = 10\cos(2t-1) - 100\ \sin(100t-30)\cos(t)$
6. $v(t) = 10\cos(2t+1)e^{t} + \sin(t) - 1/2$
7. $v(t) = \cos(2t) + 10 + \dfrac{\sin(t)}{t}$
8. $v(t) = 1/t - 10\cos(2t+1) + 10$
9. $v(t) = 1/2\cos(4t-3) - 100\ \sin(100t-30) + \cos(t/2) - 1$
10. $v(t) = 1 + 5\cos(10t-1) - 100\sin(100t-30) + \cos(7t)$

If they are periodic, find the period, T, for each signal.

EOCP 1.2

Find the energy in each of the following signals.

1. $v(t) = e^{-3t} \quad 0 \le t \le 100$
2. $v(t) = te^{-3t} \quad 0 \le t \le 10$
3. $v(t) = e^{-3t} + 1 \quad 0 \le t \le 10$
4. $v(t) = \sin(t-1) \quad -1 \le t \le 10$
5. $v(t) = \sin(t)e^{-3t} \quad 0 \le t \le 10$
6. $v(t) = \sin(t) + \cos(t) \quad 0 \le t \le 1$
7. $v(t) = 10e^{-3t} + t \quad -2 \le t \le 10$
8. $v(t) = 1 + \sin(t) \quad 0 \le t \le 10$
9. $v(t) = 10 \quad -1 \le t \le 1$
10. $v(t) = 1 - \cos(2t) \quad 0 \le t \le 1$

EOCP 1.3

Find the average power in each of the following signals, assuming a load of 1-ohm resistance.

1. $v(t) = 1 - \sin(t-1)$
2. $v(t) = \sin(t) + \cos(t)$
3. $v(t) = 1 + \sin(t)/21$

EOCP 1.4

Consider the following signal.

$$v(t) = \begin{cases} t & 0 \le t \le 1 \\ 1 & 1 \le t \le 2 \\ 0 & \text{elsewhere} \end{cases}$$

a. Mathematically, find
   1. $v(-t)$
   2. $v(-t+1)$
   3. $v(-t/2+1)$
   4. $2v(t/2-5)$
b. Sketch the above four versions of $v(t)$.

EOCP 1.5

Consider the signals shown in Figures 1.54a–1.54c. Plot and mathematically find for all signals.

1. $v(-t-1)$
2. $v(1-t)$
3. $v(-1-t/2)$
4. $v(-t/2-1/2)$

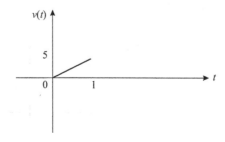

FIGURE 1.54a   Signal for EOCP 1.5.

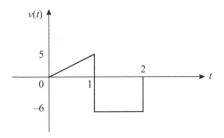

FIGURE 1.54b   Signal for EOCP 1.5.

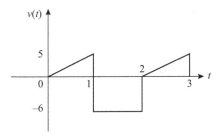

FIGURE 1.54c   Signal for EOCP 1.5.

EOCP 1.6

Express each of the signals shown in Figures 1.55–1.57 as the sum of even and odd signals.

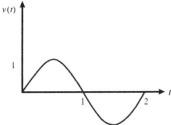

FIGURE 1.55   Signals for EOCP 1.6.

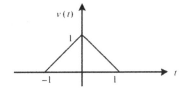

FIGURE 1.56   Signals for EOCP 1.6.

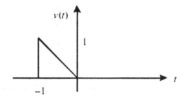

FIGURE 1.57   Signals for EOCP 1.6.

EOCP 1.7

Evaluate the following integrals using the sifting property of the delta function.

1. $\displaystyle\int_{-1}^{2}(t-\sin(t))\delta(t-1)dt$

2. $\displaystyle\int_{-1}^{2}(t-\sin(t))t\delta(t)dt$

3. $\displaystyle\int_{-1}^{2}t\delta(t/2+3)dt$

4. $\displaystyle\int_{-1}^{2}(t-\sin(t))t\delta(2t+1)dt$

5. $\displaystyle\int_{-\infty}^{2}\left(e^{-2t}\right)\delta(t-1)dt$

6. $\displaystyle\int_{-\infty}^{\infty}\left(e^{-2t}\right)\sin(t)\delta(2t-1)dt$

7. $\displaystyle\int_{-1}^{\infty}\left(e^{-2t}\right)t\sin(t)\delta(t+1/2)dt$

8. $\displaystyle\int_{-\infty}^{0}\left(e^{-2t}\right)(\sin(t)/t)\delta(t+1)dt$

9. $\displaystyle\int_{-\infty}^{2}\left(e^{-2t}+\cos(t/20)\right)\delta(t)dt$

10. $\displaystyle\int_{-\infty}^{10}\left(e^{-2t}+1\right)\delta(t-20)dt$

FIGURE 1.58   Signal for EOCP 1.8.

EOCP 1.8

Express the signal shown in Figure 1.58 as a combination of unit step signals.

EOCP 1.9

Repeat the above problem considering the signal in Figure 1.59.

EOCP 1.10

Find the derivative of the signals shown in Figures 1.60a–1.60e.

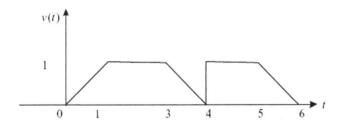

FIGURE 1.59   Signal for EOCP 1.9.

FIGURE 1.60a   Signal for EOCP 1.10.

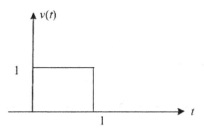

FIGURE 1.60b Signal for EOCP 1.10.

FIGURE 1.60c Signal for EOCP 1.10.

FIGURE 1.60d Signal for EOCP 1.10.

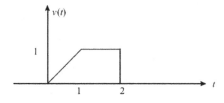

FIGURE 1.60e Signal for EOCP 1.10.

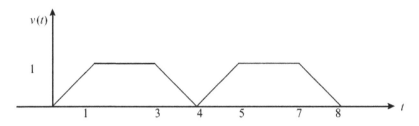

FIGURE 1.61   Signal for EOCP 1.12.

EOCP 1.11

Consider that the signal $v(t) = 10 \cos(100t - 100)$ volts represents the voltage across the terminals of a resistor and the current through the resistor is given by $i(t) = \cos(100t - 100)$ amps.

1. What is the instantaneous power consumed by the resistor?
2. Is the instantaneous power periodic?
3. If the instantaneous power signal is periodic, what is the period?

EOCP 1.12

An object of mass m acquires a velocity, as given in Figure 1.61.

1. What is the acceleration of the object? Sketch it.
2. What is the distance covered? Sketch it.

EOCP 1.13

Consider the circuit shown in Figure 1.62.
   If $x(t)$ is an impulse of strength 10 volts that is applied at $t = 0$

1. For $x(t)$ as given, what is the output $y(t)$ for $t > 0$?
2. What is the energy in $y(t)$?

FIGURE 1.62   Circuit for EOCP 1.13.

FIGURE 1.63   Circuit for EOCP 1.14.

3. What is the current in the resistor R?
4. What is the energy in the resistor's current signal?

EOCP 1.14

Consider the circuit shown in Figure 1.63.
  If $x(t)$ is an impulse that is applied at $t = 0$,

1. For the given $x(t)$, what is $y(t)$, the voltage in the capacitor if the inductance is 1 henry and the capacitance is 1 F?
2. What is the energy in $y(t)$?
3. If the capacitance is zero, what is the current in the inductor?
4. What is the energy in the inductor current?

EOCP 1.15

An object of mass M is pushed with a force F for a period of 10 seconds. Consider the two velocity signals over the same period of time, as shown in Figure 1.64.
  What is the force signal for each case? Sketch the force signals.

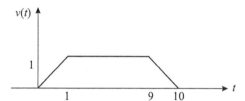

FIGURE 1.64   Signals for EOCP 1.15.

REFERENCES

Denbigh, P. *System Analysis and Signal Processing*, Reading, MA: Addison-Wesley, 1998.

Phillips, C.L. and Parr, J.M. *Signals, Systems, and Transforms*, 2nd ed., Englewood Cliffs, NJ: Prentice-Hall, 1999.

Pratap, R. *Getting Started with MATLAB 5*, New York, NY: Oxford University Press, 1999.

Strum, R.D. and Kirk, D.E. *Contemporary Linear Systems*, Boston, MA: PWS, 1996.

The MathWorks. *The Student Edition of MATLAB*, Englewood Cliffs, NJ: Prentice-Hall, 1997.

Ziemer, R.E., Tranter, W.H., and Fannin, D.R. *Signals Systems Continuous and Discrete*, 4th ed., Englewood Cliffs, NJ: Prentice-Hall, 1998.

2 Continuous Systems

2.1 DEFINITION OF A SYSTEM

A system is an assemblage of things that are combined to form a complex whole. When we study systems we study the interactions and behavior of such an assemblage when subjected to certain conditions. These conditions are called inputs. In its simplest case, a system can have one input and one output. The majority of this book will deal with linear systems. We will call the input $x(t)$ and the output $y(t)$ as depicted in Figure 2.1.

2.2 INPUT AND OUTPUT

If an input signal $x(t)$ is available at the input of the linear system shown in Figure 2.1, the system will operate on the signal $x(t)$ and produce an output signal that we call $y(t)$. As an example, consider the case of an elevator where you push a button to go to the fifth floor. Pushing the button is the input $x(t)$. The elevator is the system under consideration here. The elevator system basically consists of the small room we ride in and the motor that drives the elevator belt. The input signal $x(t)$ "asks" the elevator to move to the fifth floor. The elevator system will process this request and move to the fifth floor. The motion of the elevator to the selected floor is called the output $y(t)$.

Pushing the button in this elevator example produces an electrical signal $x(t)$. This signal drives the motor of the elevator to produce a rotational motion, which is transferred, via some gears, to a translational motion. This translational motion is the output $y(t)$. To summarize, when an electrical input signal or request $x(t)$ is applied to the elevator system, the elevator will operate on the signal and produce $y(t)$, which in this example can be thought of as a translational motion.

2.3 LINEAR CONTINUOUS SYSTEM

A linear continuous system is one where if (a) the input is $\alpha x_1(t)$, the output is $\alpha y_1(t)$; and (b) the input is $x_1(t) + x_2(t)$, the output is $y_1(t) + y_2(t)$. By combining the two conditions, a system is considered linearly continuous if for the input $\alpha x_1(t) + \beta x_2(t)$, the output is $\alpha y_1(t) + \beta y_2(t)$, where α and β are constants.

Example 2.1

Consider the voltage divider circuit shown in Figure 2.2, where $x(t)$ is the input and $y(t)$ is the output voltage across the resistor R_2.

47

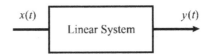

FIGURE 2.1   Linear system representation.

FIGURE 2.2   Circuit for Example 2.1.

The output voltage $y(t)$ is

$$y(t) = \left[R_2 / (R_1 + R_2) \right] x(t).$$

Is this system linear?

SOLUTION

Using the definition of linearity introduced above, we can proceed as in the following. If the input is $x_1(t)$, the output is $y_1(t)$, and therefore

$$y_1(t) = \left[R_2 / (R_1 + R_2) \right] x_1(t).$$

If the input is $x_2(t)$, the output is $y_2(t)$, and therefore

$$y_2(t) = \left[R_2 / (R_1 + R_2) \right] x_2(t).$$

If the input is $\alpha x_1(t) + \beta x_2(t)$, the output is $y(t)$, and therefore

$$y(t) = \left[R_2 / (R_1 + R_2) \right] \left(\alpha x_1(t) + \beta x_2(t) \right)$$

or

$$y(t) = \alpha \left[R_2 / (R_1 + R_2) \right] x_1(t) + \beta \left[R_2 / (R_1 + R_2) \right] x_2(t),$$

which is

$$y(t) = \alpha y_1(t) + \beta y_2(t).$$

Therefore, the system is said to be linear.

Example 2.2

Consider the following system

$$y(t) = \sqrt{x(t)}$$

Is this system linear?

SOLUTION

Consider two cases of the input signals, $x_1(t)$ and $x_2(t)$. The corresponding outputs are then given by

$$y_1(t) = \sqrt{x_1(t)}$$

for $x_1(t)$ and

$$y_2(t) = \sqrt{x_2(t)}$$

for $x_2(t)$. Now let us consider further that $\alpha x_1(t) + \beta x_2(t)$ has been applied to the system as its input. The corresponding output is then given by

$$y(t) = \sqrt{\alpha x_1(t) + \beta x_2(t)}.$$

But

$$y(t) = \sqrt{\alpha x_1(t) + \beta x_2(t)} \neq \alpha y_1(t) + \beta y_2(t) = \alpha \sqrt{x_1(t)} + \beta \sqrt{x_2(t)}.$$

Therefore, the system is nonlinear.

Example 2.3

Consider the following system

$$y(t) = \left(2/\left[2x(t) + 1\right]\right) x(t)$$

Is this system linear?

SOLUTION

As in Example 2.2, consider two cases of the input signals, $x_1(t)$ and $x_2(t)$. The corresponding outputs are then given by

$$y_1(t) = \left(2/\left[2x_1(t) + 1\right]\right) x_1(t)$$

and

$$y_2(t) = \left(2/\left[2x_2(t) + 1\right]\right) x_2(t).$$

Now let us apply $\alpha x_1(t) + \beta x_2(t)$ to the system as its input. The corresponding output is then given by

$$y(t) = \left(2/\left[2\left(\alpha x_1(t) + \beta x_2(t)\right) + 1\right]\right)\left(\alpha x_1(t) + \beta x_2(t)\right)$$

or

$$y(t) = \frac{\left[2\alpha x_1(t) + 2\beta x_2(t)\right]}{\left[2\alpha x_1(t) + 2\beta x_2(t) + 1\right]}$$

$$\neq \alpha y_1(t) + \beta y_2(t) = \left[2\alpha x_1(t)\right]/\left[2x_1(t) + 1\right] + \left[2\beta x_2(t)\right]/\left[2x_2(t) + 1\right].$$

Therefore, the system is nonlinear.

2.4   TIME-INVARIANT SYSTEM

A system is said to be time-invariant if for a shifted input $x(t - t_0)$, the output of the system is $y(t - t_0)$.

To see if a system is time-invariant or time-variant we do the following:

1. Find the output $y_1(t - t_0)$ that corresponds to the input $x_1(t)$.
2. Let $x_2(t) = x_1(t - t_0)$ and then find the corresponding output $y_2(t)$.
3. Check if $y_1(t - t_0) = y_2(t)$. If this is true then the system is time-invariant. Otherwise it is time-variant.

Example 2.4

Let

$$y(t) = \cos(x(t)).$$

Find out if the system is time-variant or time-invariant.

SOLUTION

According to the definition of time-invariant systems, we write

$$y_1(t) = \cos(x_1(t)).$$

$y_1(t)$ shifted by t_0 is

$$y_1(t - t_0) = \cos(x_1(t - t_0)).$$

Also

$$y_2(t) = \cos(x_1(t - t_0)).$$

Since

$$y_1(t - t_0) = y_2(t)$$

the system is time-invariant.

Example 2.5

Let

$$y(t) = x(t)\cos(t).$$

Find out if the system is time-variant or time-invariant.

SOLUTION

STEP 1

$$y_1(t) = x_1(t)\cos(t)$$

Therefore,

$$y_1(t - t_0) = x_1(t - t_0)\cos(t - t_0).$$

STEP 2

$$y_2(t) = x_1(t - t_0)\cos(t)$$

STEP 3

$$y_1(t - t_0) \neq y_2(t)$$

Therefore, the system is time-variant.

Example 2.6

Let

$$y(t) = te^{-t}x(t).$$

Find out if the system is time-variant or time-invariant.

SOLUTION

STEP 1

$$y_1(t) = te^{-t}x_1(t)$$

Therefore,

$$y_1(t-t_0) = (t-t_0)e^{-(t-t_0)}x_1(t-t_0).$$

STEP 2

$$y_2(t) = te^{-t}x_1(t-t_0)$$

STEP 3

$$y_1(t-t_0) \neq y_2(t)$$

Therefore, the system is time-variant.

2.5 SYSTEMS WITHOUT MEMORY

Consider the system described by the following integral

$$y(t) = \int_{-\infty}^{t} x(\tau)d\tau.$$

The value of $y(t)$, at a particular time t, depends on the time from $-\infty$ to t. In this case, we say that the system has a memory. On the other hand, consider another system given by

$$y(t) = cx(t)$$

where c is a constant.

At any value of time t, $y(t)$ depends totally on $x(t)$ at that point in time. In a case such as this we say the system is without a memory.

If the system is without memory we can write $y(t)$ as a function of $x(t)$ as

$$y(t) = F(x(t)).$$

2.6 CAUSAL SYSTEMS

A causal system is a system where the output $y(t)$ at a certain time t_1 depends on the input $x(t)$ for $t < t_1$.

Example 2.7

Let $x(t)$ be an input as shown in Figure 2.3.

Let $y(t) = x(t + 1)$. Is the system causal or noncausal? What happens to the system if $y(t) = x(t - 1)$?

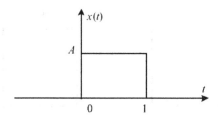

FIGURE 2.3 Signal for Example 2.7.

SOLUTION

1. Let us look at the graph for the system

$$y(t) = x(t+1)$$

as shown in Figure 2.4.

You can observe that output $y(t)$ appears before the input has been introduced. Therefore, the given system is a noncausal system.

2. Let us look at the graph shown in Figure 2.5 for the system

$$y(t) = x(t-1).$$

You can observe in this case that the output $y(t)$ appears only after the input has been introduced. Therefore, the given system is causal.

FIGURE 2.4 Signal for Example 2.7.

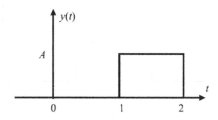

FIGURE 2.5 Signal for Example 2.7.

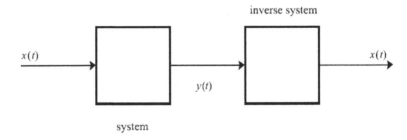

FIGURE 2.6 Inverse system.

2.7 THE INVERSE OF A SYSTEM

If we can determine the input by measuring the output, then the system under consideration is said to be invertible.

Pictorially, an inverse system can be represented as shown in Figure 2.6. Notice that if two inputs give the same output then the system is not invertible.

Example 2.8

Consider the following systems.

    1. $y(t) = x(t)$
    2. $y(t) = 2x(t)$
    3. $y(t) = a \cos(x(t))$
    4. $y(t) = \displaystyle\int_{-\infty}^{t} x(\tau)d\tau$ such that $y(-\infty) = 0$

Are these systems invertible?

SOLUTION

The first system is invertible. The corresponding pictorial representation is shown in Figure 2.7.

The second system is also invertible. Figure 2.8 shows the corresponding pictorial depiction.

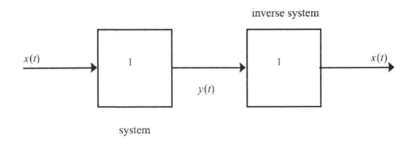

FIGURE 2.7 System for Example 2.8.

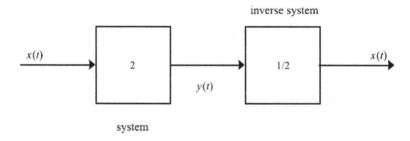

FIGURE 2.8   System for Example 2.8.

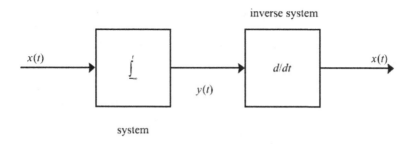

FIGURE 2.9   System for Example 2.8.

The third system is not invertible. Why?

Let us consider two inputs, $x_1(t) = x(t)$ and $x_2(t) = x(t) + n\pi$, where n is an even integer.

For this system the output corresponding to $x_1(t) = x(t)$ is given by

$$y_1(t) = a \, \cos(x(t)),$$

and the output corresponding to $x_2(t) = x(t) + n\pi$ is given by

$$y_2(t) = a \, \cos(x(t) + n\pi) = a \, \cos(x(t)).$$

We can see here that two different inputs produced the same output. Therefore, the system is not invertible as claimed.

The fourth system is invertible. Notice that $\dfrac{d}{dt} y(t) = x(t)$ as shown in Figure 2.9.

2.8   STABLE SYSTEMS

The signal $x(t)$ is considered bounded if $|x(t)| < \beta < \infty$ for all t, where β is a real number. A system is said to be BIBO (bounded-input bounded-output) stable if and when the input is bounded the output is also bounded. The output $y(t)$ is bounded if $|y(t)| < \beta < \infty$.

Example 2.9

Are the following systems bounded if it is known that the input $x(t)$ is bounded?

1. $y(t) = e^{x(t)}$
2. $y(t) = \int_{-\infty}^{t} x(\tau) d\tau$

SOLUTION

1. Since $x(t)$ is bounded, then $|x(t)| < \beta < \infty$. Accordingly, the magnitude of the output is

$$|y(t)| = \left|e^{x(t)}\right| \le \left|e^{\beta}\right| = e^{\beta} < \infty.$$

Therefore, the system is BIBO stable.

2. Since

$$y(t) = \int_{-\infty}^{t} x(\tau) d\tau,$$

the magnitude of the output is

$$|y(t)| = \left|\int_{-\infty}^{t} x(\tau) d\tau\right| \le \int_{-\infty}^{t} |x(\tau)| \, d\tau \le \beta \int_{-\infty}^{t} d\tau.$$

Notice that

$$\int_{-\infty}^{t} d\tau = \tau \big|_{-\infty}^{t} = t + \infty = \infty.$$

Therefore, the system is not stable.

2.9   CONVOLUTION

As discussed earlier, by using the sifting property of the impulse signal we can write

$$\int_{t_1}^{t_2} x(\tau)\delta(\tau) d\tau = x(0) \text{ for } t_1 < 0 < t_2$$

Also

$$\int_{-\infty}^{+\infty} x(\tau)\delta(t - \tau) d\tau = x(t) \text{ for } -\infty < t < \infty$$

This last integral indicates that any signal can be represented as a sum of weighted impulses. For linear systems, if $x_1(t)$, $x_2(t)$, $x_3(t)$, $x_4(t)$, ..., $x_n(t)$ are the system inputs, the output of the system is obtained by superposition. It will be the sum of all outputs resulting from the individual inputs where each is presented alone. Mathematically we can write

$$y(t) = y_1(t) + y_2(t) + y_3(t) + y_4(t) + \ldots + y_n(t)$$

where $y(t)$ is the output due to the input $x(t)$.

The convolution integral is given by

$$y(t) = \int_{-\infty}^{+\infty} x(\tau) h(t, \tau) d\tau,$$

where $h(t - \tau)$ is the output at time t for the input $\delta(t - \tau)$ applied at time $t = \tau$. If the system is time-invariant the convolution integral simplifies to

$$y(t) = \int_{-\infty}^{+\infty} x(\tau) h(t - \tau) d\tau, \tag{2.1}$$

where $h(t)$ is the impulse response of the linear time-invariant (LTI) system and represents the output of the system at time t to a unit impulse signal when the system is relaxed (i.e., no initial conditions). As a matter of notation we write the equation above as

$$y(t) = x(t) * h(t).$$

We read this equation as $y(t)$ equals $x(t)$ convolved with $h(t)$.

Example 2.10

What kinds of systems are the following?

1. $x(t) * \delta(t)$
2. $x(t) * u(t)$
3. $x(t) * \dfrac{d}{dt} \delta(t)$

SOLUTION

For the first system

$$x(t) * \delta(t) = \int_{-\infty}^{+\infty} x(\tau) \delta(t - \tau) d\tau = x(t).$$

In the above equation $h(t) = \delta(t)$ and we say the system is a unity system. For the second system

$$x(t) * u(t) = \int_{-\infty}^{+\infty} x(\tau) u(t-\tau) d\tau = \int_{-\infty}^{t} x(\tau) d\tau.$$

In the above equation $h(t) = u(t)$ and the system is an integrator. For the third system

$$x(t) * \frac{d}{dt}\delta(t) = \int_{-\infty}^{+\infty} x(\tau) \frac{d}{dt}\delta(t-\tau) d\tau = \frac{d}{dt} x(t).$$

In the above equation $h(t) = \dfrac{d}{dt}\delta(t)$ and the system is a differentiator.

2.10   SIMPLE BLOCK DIAGRAMS

We have called $h(t)$ the impulse response of the system under consideration.

1. A block diagram representation for a system with $x(t)$ as the input, $y(t)$ as the output, and $h(t)$ as the impulse response can be represented as shown in Figure 2.10. Mathematically we write

$$y(t) = x(t) * h(t).$$

2. If we have two systems connected in series as shown in Figure 2.11, then mathematically we write

$$z(t) = x(t) * h_1(t)$$

and the output is

$$y(t) = z(t) * h_2(t).$$

FIGURE 2.10   General block diagram representation.

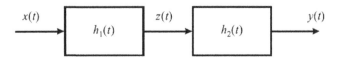

FIGURE 2.11   Series system.

Therefore

$$y(t) = \left[\left(x(t) * h_1(t) \right) * h_2(t) \right].$$

This means that we convolve $x(t)$ with $h_1(t)$ to get $z(t)$, and then convolve $z(t)$ in turn with $h_2(t)$ to obtain $y(t)$.

3. If we have two systems connected in parallel as shown in Figure 2.12, then mathematically we write

$$z_1(t) = x(t) * h_1(t)$$

and

$$z_2(t) = x(t) * h_2(t).$$

The output $y(t)$ is

$$y(t) = z_1(t) + z_2(t).$$

Substituting for $z_1(t)$ and $z_2(t)$ we get

$$y(t) = \left[x(t) * h_1(t) \right] + \left[x(t) * h_2(t) \right].$$

Important properties:

1. Multiplication of the signal $x(t)$ by the shifted impulse signal

$$x(t)\delta(t-a) = x(a)\delta(t-a)$$

2. The sifting property

$$\int_{-\infty}^{\infty} x(t)\delta(t-a)\,dt = x(a)$$

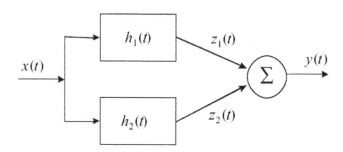

FIGURE 2.12 Parallel system.

3. The convolution property

$$x(t) * \delta(t-a) = x(t-a)$$

Example 2.11

Let the impulse response for an LTI system be

$$h(t) = e^{-at}u(t)$$

where a is a positive constant. If the input to the system is $x(t) = u(t)$, find the output $y(t)$ using the convolution integral.

SOLUTION

We can use the convolution integral to find the output $y(t)$ as

$$y(t) \int_{-\infty}^{\infty} x(\tau) h(t-\tau) d\tau.$$

Substituting for $x(t)$ and $h(t)$ we get

$$y(t) \int_{-\infty}^{\infty} u(\tau) h(t-\tau) d\tau,$$

which simplifies to

$$y(t) \int_{-\infty}^{\infty} u(\tau) e^{-a(t-\tau)} u(t-\tau) d\tau.$$

As we see in Figure 2.13

$$u(\tau)u(t-\tau) = 0 \text{ for } t < 0$$

and

$$u(\tau)u(t-\tau) = 1 \text{ for } t > 0.$$

Therefore, the above convolution equation reduces to

$$y(t) = \int_{0}^{t} e^{-a(t-\tau)} \, d\tau = e^{-at} \int_{0}^{t} e^{a\tau} d\tau.$$

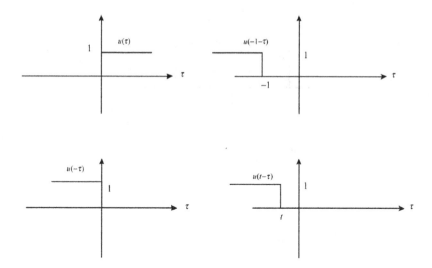

FIGURE 2.13   Signals for Example 2.11.

Finally

$$y(t) = \frac{1}{a}\left[1 - e^{-at}\right] \text{ for } t > 0.$$

2.11   GRAPHICAL CONVOLUTION

Graphical convolution is best explained by an example.

Example 2.12

Let $x(t)$ and $h(t)$ be given as in Figure 2.14. Find $y(t)$ using graphical convolution.

SOLUTION

STEP 1

Fix (but not slide) either $x(t)$ or $h(t)$ on the τ axis. Here we will fix $h(t)$.

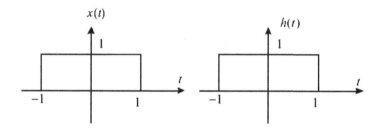

FIGURE 2.14   Signals for Example 2.12.

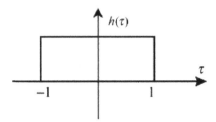

FIGURE 2.15   Signal for Example 2.12.

STEP 2

Sketch $h(\tau)$ by first writing the mathematical equation for $h(t)$ as

$$h(t) = \begin{cases} 1 & -1 < t < 1 \\ 0 & \text{otherwise} \end{cases}$$

and

$$h(\tau) = \begin{cases} 1 & -1 < \tau < 1 \\ 0 & \text{otherwise.} \end{cases}$$

The sketch is given in Figure 2.15.

STEP 3

We will sketch $x(t - \tau)$ by first writing its mathematical representation as

$$x(t - \tau) = \begin{cases} 1 & -1 < t - \tau < 1 \\ 0 & \text{otherwise.} \end{cases}$$

Fixing the time axis we get

$$x(t - \tau) = \begin{cases} 1 & -1 - t < -\tau < -t + 1 \\ 0 & \text{otherwise} \end{cases}$$

$$x(t - \tau) = \begin{cases} 1 & 1 + t > \tau > t - 1 \\ 0 & \text{otherwise} \end{cases}$$

and finally

$$x(t - \tau) = \begin{cases} 1 & t - 1 < \tau < t + 1 \\ 0 & \text{otherwise.} \end{cases}$$

The sketch is given in Figure 2.16.

STEP 4

Move $x(t - \tau)$ to the left of $h(\tau)$ first and then start moving it (sliding $x(t - \tau)$) to the right on the τ axis. This is shown in Figure 2.17.

FIGURE 2.16 Signal for Example 2.12.

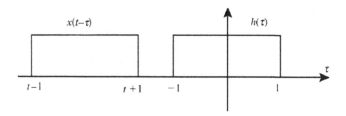

FIGURE 2.17 Signal for Example 2.12.

Now start moving $x(t - \tau)$ to the right to produce overlapping with $h(\tau)$ and determine the corresponding output.

1. When $t = -2$, we have Figure 2.18. The two functions are just overlapping and therefore

$$y(t) = 0 \text{ for } t < -2.$$

2. When $-2 \le t < -1$ we have the picture shown in Figure 2.19, where the X in the figure indicates the overlapping region.

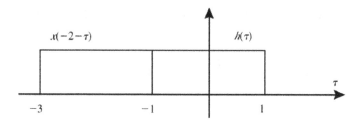

FIGURE 2.18 Signal for Example 2.12.

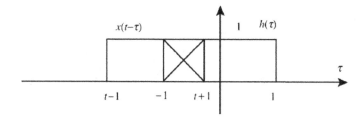

FIGURE 2.19 Signal for Example 2.12.

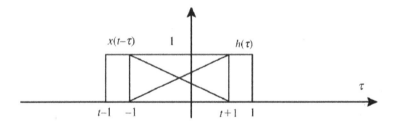

FIGURE 2.20   Signal for Example 2.12.

The limits for the corresponding convolution integral are the limits you see at the overlapping region and the magnitudes of the two functions are 1 and 1, respectively. Therefore

$$y(t) = \int_{-1}^{t+1} (1)(1)\,d\tau = (t+1) - (-1)$$

$$y(t) = t + 2 \text{ for } -2 \le t < -1.$$

3. When $-1 \le t < 0$ we have the picture shown in Figure 2.20.
   The overlapping region is stretched now from -1 to $t + 1$. Therefore

$$y(t) = \int_{-1}^{t+1} (1)(1)\,d\tau = (t+1) - (-1)$$

$$y(t) = t + 2 \text{ for } -1 \le t < 0.$$

4. When $0 \le t < 1$ we have the picture shown in Figure 2.21. The overlapping extends now from $t - 1$ to 1. Therefore

$$y(t) = \int_{t-1}^{1} (1)(1)\,d\tau = (1) - (t-1)$$

$$y(t) = -t + 2 \text{ for } 0 \le t < 1.$$

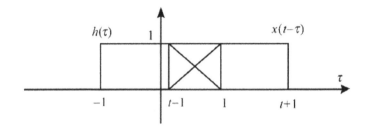

FIGURE 2.21   Signal for Example 2.12.

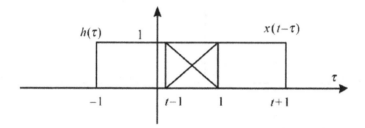

FIGURE 2.22 Signal for Example 2.12.

5. When $1 \leq t < 2$ we have the picture shown in Figure 2.22, where in this case

$$y(t) = \int_{t-1}^{1} (1)(1) \, d\tau = (1) - (t-1)$$

$$y(t) = -t + 2 \text{ for } 1 \leq t < 2.$$

6. When $2 \leq t < 3$ we have the picture shown in Figure 2.23.

In this case, we see no overlap between the two graphs. Therefore

$$y(t) = 0 \text{ for } 2 \leq t < 3.$$

Finally we can write the final results as

$$y(t) = \begin{cases} 0 & t < -2 \\ t+2 & -2 \leq t \leq 0 \\ -t+2 & 0 \leq t \leq 2 \\ 0 & t > 2. \end{cases}$$

The output $y(t) = x(t) * h(t)$ thus takes the graphical form such as that shown in Figure 2.24.

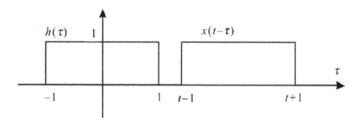

FIGURE 2.23 Signal for Example 2.12.

FIGURE 2.24   Signal for Example 2.12.

2.12   DIFFERENTIAL EQUATIONS AND PHYSICAL SYSTEMS

In most physical systems, $y(t)$ is a solution of a differential equation. Since we are dealing with LTI systems, our differential equations will be linear and have constant coefficients. The general form is

$$a_n \frac{d^n}{dt} y(t) + a_{n-1} \frac{d^{n-1}}{dt} y(t) + a_{n-2} \frac{d^{n-2}}{dt} y(t) + \cdots + a_0 y(t)$$

$$= b_m \frac{d^m}{dt} x(t) + b_{m-1} \frac{d^{m-1}}{dt} x(t) + \cdots + b_0 x(t) \tag{2.2}$$

where $x(t)$ is the input and $y(t)$ is the output of the system under consideration. In order to solve for the output $y(t)$, we must have all initial conditions needed. We will need n initial conditions, where n is the order of the differential equation. These initial conditions are

$$\frac{d^{n-1}}{dt} y(0), \frac{d^{n-2}}{dt} y(0), \ldots, y(0). \tag{2.3}$$

2.13   HOMOGENEOUS DIFFERENTIAL EQUATIONS AND THEIR SOLUTIONS

If we set the right-hand side of Equation (2.2) to zero, we will have the homogeneous differential equation

$$a_n \frac{d^n}{dt} y(t) + a_{n-1} \frac{d^{n-1}}{dt} y(t) + a_{n-2} \frac{d^{n-2}}{dt} y(t) + \ldots + a_0 y(t) = 0 \tag{2.4}$$

with the initial conditions

$$\frac{d^{n-1}}{dt} y(0), \frac{d^{n-2}}{dt} y(0), \ldots, y(0).$$

To solve the homogeneous differential equation, we form the characteristic algebraic equation

$$a_n m^n + a_{n-1} m^{n-1} + \ldots + a_0 = 0. \tag{2.5}$$

The roots for this algebraic equation are called the poles or the eigenvalues of the system.

2.13.1 CASE WHEN THE ROOTS ARE ALL DISTINCT

If the roots are all distinct, the solution of the homogeneous equation is

$$y(t) = c_1 e^{m_1 t} + c_2 e^{m_2 t} + \ldots + c_n e^{m_n t} \tag{2.6}$$

where m_1, m_2, ..., m_n are the n distinct roots of the characteristic equation describing the system. The constants are determined using the initial conditions, which are usually given along with the description of the system.

2.13.2 CASE WHEN TWO ROOTS ARE REAL AND EQUAL

If we have a second-order differential equation, where m_1 and m_2 are real and equal, we are then tempted to write the solution as

$$y(t) = c_1 e^{m_1 t} + c_2 e^{m_2 t} = c_1 e^{m_1 t} + c_2 e^{m_1 t} = (c_1 + c_2) e^{m_1 t} = c e^{m_1 t}. \tag{2.7}$$

Notice that this is a solution of a first-order system with one energy storage element. In this case, the solution is corrected and written as

$$y(t) = c_1 e^{m_1 t} + c_2 t e^{m_1 t}$$

where the two terms are independent. If the differential equation is of third-order type, with m_3 as another root, then the solution is

$$y(t) = c_1 e^{m_1 t} + c_2 t e^{m_2 t} + c_3 e^{m_3 t}.$$

2.13.3 CASE WHEN TWO ROOTS ARE COMPLEX

If m_1 and m_2 are two complex roots of the characteristic equation, the two roots should appear in a complex conjugate pair. In this case, the roots are still distinct but complex. The solution is still of the form

$$y(t) = c_1 e^{m_1 t} + c_2 e^{m_2 t}.$$

In general, we can write m_1 and m_2 as $a + jb$ and $a - jb$. The solution becomes

$$y(t) = c_1 e^{(a+jb)t} + c_2 e^{(a-jb)t}.$$

Using the fact that

$$e^{\pm jx} = \cos \pm j \sin x$$

and

$$e^{x+y} = e^x e^y,$$

we can simplify the above equation as

$$y(t) = c_1 e^{(a+jb)t} + c_2 e^{(a-jb)t}$$

$$y(t) = c_1 e^{at} \left[\cos bt + j \sin bt \right] + c_2 e^{at} \left[\cos bt - j \sin bt \right].$$

Simplifying we get

$$y(t) = e^{at} \left[(c_1 + c_2) \cos bt + (c_1 + c_2) j \sin bt \right].$$

If we let $c_1 + c_2 = a_1$ and $(c_1 - c_2)j = a_2$ then

$$y(t) = e^{at} \left[a_1 \cos bt + a_2 \sin bt \right] \tag{2.8}$$

where a_1 and a_2 can be found using the given initial conditions.

2.14 NONHOMOGENEOUS DIFFERENTIAL EQUATIONS AND THEIR SOLUTIONS

A differential equation is referred to as nonhomogeneous if it has an input (a forcing function) that appears to the right of the equal sign as in the following

$$a_n \frac{d^n}{dt} y(t) + a_{n-1} \frac{d^{n-1}}{dt} y(t) + a_{n-2} \frac{d^{n-2}}{dt} y(t) + \cdots + a_0 y(t) = x(t). \tag{2.9}$$

A solution of this equation has the form

$$y(t) = y_h(t) + y_p(t) \tag{2.10}$$

where $y_h(t)$ is the homogenous part of the solution and $y_p(t)$ is the particular part. The homogenous part has the form

$$y_h(t) = c_1 e^{m_1 t} + c_2 e^{m_2 t} + \ldots + c_n e^{m_n t}.$$

In the homogenous solution we cannot use the initial conditions

$$\frac{d^{n-1}}{dt}y(0), \frac{d^{n-2}}{dt}y(0),...,y(0)$$

to find the constants. These initial conditions should be applied to $y(t)$ not $y_h(t)$. After we find the combined solution

$$y(t) = y_h(t) + y_p(t)$$

we can use the initial condition to find the cs.

2.14.1 How Do We Find the Particular Solution?

Table 2.1 will help us in finding the form of the particular solution. The constants that correspond to the particular solution can be obtained by simply substituting $y_p(t)$ in the given nonhomogeneous equation and then equating coefficients. The constants corresponding to the homogeneous equation are determined using the initial conditions that come with the given nonhomogeneous differential equation with its constants, after the two solutions, $y_h(t)$ and $y_p(t)$, have been determined. These two solutions are added together to form the overall solution $y(t)$.

Example 2.13

Consider the system described by the following differential equation

$$\frac{d^2}{dt}y(t) + \frac{d}{dt}y(t) = 2t$$

with the initial conditions

$$y(0) = 0 \text{ and } \frac{d}{dt}y(0) = 0.$$

What is the solution $y(t)$ (the output)?

TABLE 2.1
Particular Solution Forms

Input	Particular Solution
constant A	constant B
Ae^{at}	Be^{at}
$A \cos (at)$ or $A \sin (at)$	$B_1\cos (at) + B_2\sin (at)$
At^n	$B_nt^n + B_{n-1}t^{n-1} + \cdots + B_0$

SOLUTION

$y(t)$ will have two components: the homogeneous solution and the particular solution.

We find the homogeneous solution by first finding the characteristic equation as

$$m^2 + m = m(m+1) = 0.$$

The two roots are real and distinct: $m = 0$ and $m = -1$. The corresponding homogeneous solution is therefore

$$y_h(t) = c_1 e^{-1t} + c_2 e^{0t} = c_1 e^{-t} + c_2.$$

Here, the input $x(t)$ is $2t$. If we look at Table 2.1, we find out that the form of the particular solution is

$$y_p(t) = B_2 t + B_1$$

If we substitute this particular solution in the given differential equation we will get

$$0 + B_2 = 2t.$$

By equating coefficients of $t(0t^0$ is compared with $B_2 t^0)$, B_2 is 0 and there is no way of calculating B_1. The terms in $y_p(t) + y_h(t)$ must be independent. Here we have the following combination as the total solution

$$y_h(t) + y_p(t) = c_1 e^{-t} + c_2 + B_2 t + B_1,$$

where c_2 and B_1 are dependents. In this situation we multiply B_1, the part of the particular solution that causes the dependency, by the minimum integer positive power of t to make the whole combination independent. This can be achieved by multiplying B_1 by t^2.

Therefore, we now rewrite the particular solution as

$$y_p(t) = B_2 t + B_1 t^2.$$

If we substitute in the differential equation given we will have

$$2B_1 + B_2 + 2B_1 t = 2t.$$

By equating coefficients, we have

$$B_1 = 1$$

and

$$2B_1 + B_2 = 0$$

which gives

$$B_2 = -2$$

The particular solution is then

$$y_p(t) = -2t + t^2.$$

The overall solution is

$$y(t) = y_h(t) + y_p(t) = c_1 e^{-t} + c_2 + -2t + t^2.$$

Using the initial conditions, we can now find the constants as in the following.

$$y(0) = 0 = c_1 + c_2$$

$$\frac{d}{dt} y(t) = -c_1 e^{-t} - 2 + 2t$$

$$\frac{d}{dt} y(0) = -c_1 - 2 = 0$$

This indicates that $c_1 = -2$ and $c_2 = 2$. The final solution is

$$y(t) = y_h(t) + y_p(t) = -2e^{-t} + 2 - 2t + t^2.$$

Example 2.14

Consider the system described by the following differential equation

$$\frac{d^3}{dt} y(t) - y(t) = \sin t$$

with initial conditions

$$y(0) = 0, \ \frac{d}{dt} y(0) = 0, \text{ and } \frac{d^2}{dt} y(0) = 0.$$

What is the output of the system?

SOLUTION

The auxiliary equation is

$$m^3 - 1 = 0$$

with the roots $m_1 = 1$, $m_2 = -0.5 + 0.866j$, and $m_3 = -0.5 - 0.866j$.

The homogeneous solution is therefore

$$y_h(t) = c_1e^t + e^{-0.5t}\left[a_1\cos(0.866t) + a_2\sin(0.866t)\right].$$

To find the particular part of the solution, we look at the input to the system. The input is sin(t), and therefore, the form of the particular solution using Table 2.1 is

$$y_p(t) = B_1\cos t + B_2\sin t.$$

Next we will find B_1 and B_2 by substituting the particular solution in the given differential equation to get

$$B_1\sin t - B_2\cos t - B_1\cos t - B_2\sin t = \sin t.$$

By grouping like terms we arrive at

$$(B_1 - B_2)\sin t - (B_2 - B_1)\cos t = \sin t,$$

and by equating coefficients we will have

$$B_1 - B_2 = 1$$

and

$$B_2 - B_1 = 0$$

where $B_1 = 1/2$ and $B_2 = -1/2$.
   The total solution now is

$$y(t) = c_1e^t + e^{-0.5t}\left[a_1\cos(0.866t) + a_2\sin(0.866t)\right] + 1/2\cos t - 1/2\sin t.$$

Using the initial conditions given, we will get three algebraic equations in three unknowns as

$$c_1 + a_1 + 0a_2 = -0.5$$

$$c_1 - 0.5a_1 + 0.886a_2 = 0.5$$

$$c_1 - 0.535a_1 - 0.886 = 0.5$$

The solution to these equations is

$$c_1 = 0.159, \ a_1 = -0.659, \text{ and } a_2 = 0.013.$$

The final solution is

$$(t) = 0.159e^t + e^{-0.5t}\left[-0.659\ \cos(0.866t) + 0.013\sin(0.866t)\right]$$

$$+1/2\cos t - 1/2\sin t.$$

2.15   THE STABILITY OF LINEAR CONTINUOUS SYSTEMS: THE CHARACTERISTIC EQUATION

If the coefficients of the characteristic equation have the same sign, then the system is stable. If there are sign changes in the coefficients of the characteristic equation, we can test stability by finding the roots of the characteristic equation analytically, or we can use the MATLAB function "roots" to find the roots of the characteristic equation. We can also find the solution to the given differential equation with zero input and nonzero initial conditions, and if the output settles at a constant value, then we conclude that the system is stable. If the coefficients in the characteristic equation are not constants, then we can use the Routh test to check for stability. The Routh test can be used for any characteristic equation whether the coefficients are constants or not. In general, consider the following characteristic equation

$$a_5 m^5 + a_4 m^4 + a_3 m^3 + a_2 m^2 + a_1 m + a_0 = 0. \tag{2.11}$$

We form the Routh array as in the following:

m^5	a_5	a_3	a_1
m^4	a_4	a_2	a_0
m^3	$(a_4 a_3 - a_5 a_2)/a_4$ call this term x_1	$(a_4 a_1 - a_5 a_0)/a_4$ call this term x_2	0
m^2	$(x_1 a_2 - a_4 x_2)/x_1$ call this term x_3	$(x_1 a_0 - a_4 (0))/x_1 = a_0$	0
m^1	$(x_3 x_2 - x_1 a_0)/x_3$ call this term x_4	$(x_3(0) - x_1(0))/x_3 = 0$	0
m^0	$(x_4 a_0 - x_3 (0))/x_4 = a_0$	$(x_4(0) - x_3(0))/x_4 = 0$	0

Now we look at the first column of the Routh array and see that every term in this first column must be positive for stability. So for stability, a_5, a_4, x_1, x_3, x_4, and a_0 must all be positive.

Example 2.15

Consider the following systems.

1. $\dfrac{d^2}{dt} y + \dfrac{d}{dt} y + y = x(t)$

2. $\dfrac{d^2}{dt} y + \dfrac{d}{dt} y - y = x(t)$

3. $\dfrac{d^2}{dt} y + \dfrac{d}{dt} y + ky = x(t)$

4. $\dfrac{d^2}{dt} y + k\dfrac{d}{dt} y + y = x(t)$

5. $\dfrac{d^2}{dt} y + \dfrac{d}{dt} y + (1 + k)y = x(t)$

6. $\dfrac{d^2}{dt} y + k\dfrac{d}{dt} y + y = x(t)$

Are the above systems stable?

SOLUTION

SYSTEM 1

The characteristic equation is

$$m^2 + m + 1 = 0.$$

We form the Routh array as

$$
\begin{array}{cc}
1 & 1 \\
1 & 0
\end{array}
$$

$$(1(1) - 1(0))/1 = 1$$

The terms in the first column are all positive; therefore, the system is stable.

SYSTEM 2

The characteristic equation is

$$m^2 + m - 1 = 0.$$

We form the Routh array as

$$
\begin{array}{cc}
1 & -1 \\
1 & 0
\end{array}
$$

$$(1(-1) - 1(0))/1 = -1$$

One term in the first column of the array is negative; therefore, the system is unstable.

SYSTEM 3

The characteristic equation is

$$m^2 + m + k = 0.$$

The Routh array is

$$
\begin{array}{cc}
1 & k \\
1 & 0
\end{array}
$$

$$(1(k) - 1(0))/1 = k$$

All terms in the first column of the array are positive if k is positive. This is to say that the system is stable for all values of $k > 0$.

SYSTEM 4

The characteristic equation is

$$m^2 + km + 1 = 0.$$

We form the Routh array as

$$
\begin{array}{cc}
1 & 1 \\
k & 0
\end{array}
$$

$$\left(k(1) - 1(0)\right)/k = 1$$

Here also k must be greater than zero for stability.

SYSTEM 5

The characteristic equation is

$$m^2 + m + (1 + k) = 0.$$

The Routh array is

$$
\begin{array}{cc}
1 & 1+k \\
1 & 0
\end{array}
$$

$$\left[1(1+k) - 1(0)\right]/1 = 1 + k$$

For stability, $(k + 1)$ should be positive; therefore, the system is stable if $k > -1$.

SYSTEM 6

The characteristic equation is

$$m^3 + km + 1 = 0.$$

The Routh array is

$$
\begin{array}{cc}
1 & k \\
0 & 1
\end{array}
$$

Before we continue and calculate the other terms in the Routh array, and since we have a leading zero in the first column, we substitute that zero with a very small positive number and we call it e. The Routh array becomes

$$
\begin{array}{cc}
1 & k \\
e & 1
\end{array}
$$

$$\left[e(k) - 1\right]/e = k - 1/e$$

For stability, $k - 1/e$ should be positive. This is equivalent to saying that $k > 1/e$. Notice that e is a very small positive number. So what is the range for k that must make this system stable? In theory we say $k > \infty$.

Use the Matlab script, Example 215. When we run this script the result is

```
enter a value for k, a negative number to exit: 1000
roots_of_system_6 =
0.0005 + 31.6228i
0.0005 - 31.6228i
- 0.0010
enter a value for k, a negative number to exit: 10000
roots_of_system_6 =
1.0e + 002*
0.0000 + 1.0000i
0.0000 - 1.0000i
0.0000
enter a value for k, a negative number to exit: 1000000
roots_of_system_6 =
1.0e + 003*
0.0000 + 1.0000i
0.0000 - 1.0000i
0.0000
enter a value for k, a negative number to exit: 10000000000
roots_of_system_6 =
1.0e + 005*
0.0000 + 1.0000i
0.0000 - 1.0000i
0.0000
enter a value for k, a negative number to exit: 10000000000000000
roots_of_system_6 =
1.0e + 008*
0.0000 + 1.0000i
0.0000 - 1.0000i
0.0000
enter a value for k, a negative number to exit: -1
```

Notice that no matter how big k is, the real parts of the roots are either positive or zero, indicating instability.

2.16   BLOCK DIAGRAM REPRESENTATION OF LINEAR SYSTEMS

Block diagrams are representations of physical systems using blocks. Individual blocks can be put together to represent the physical system in block diagram form. Individual blocks are considered next.

2.16.1   INTEGRATOR

The integrator block diagram is shown in Figure 2.25.

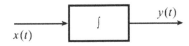

FIGURE 2.25   An integrator block diagram.

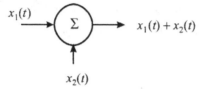

FIGURE 2.26 An adder block diagram.

The output $y(t)$ in this case is

$$y(t) = \int x(t)\,dt.$$

2.16.2 ADDER

The adder block diagram is shown in Figure 2.26.
The output $y(t)$ in this case is

$$y(t) = x_1(t) + x_2(t).$$

2.16.3 SUBTRACTOR

The subtractor block diagram is shown in Figure 2.27.
The output is

$$y(t) = x_1(t) - x_2(t).$$

2.16.4 MULTIPLIER

The multiplier block diagram is shown in Figure 2.28.
The output is

$$y(t) = k\,x(t).$$

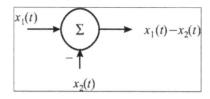

FIGURE 2.27 A subtractor block diagram.

FIGURE 2.28 A multiplier block diagram.

2.17   FROM BLOCK DIAGRAMS TO DIFFERENTIAL EQUATIONS

This process is best understood by using examples.

Example 2.16

Consider the block diagram in Figure 2.29.

The arrows indicate the direction of the signal flow. What is the system described by this block diagram? Or what is the differential equation that this system represents?

SOLUTION

Since $y(t)$ is the output of the integrator, the input to the integrator is $\dfrac{d}{dt}y(t)$. The output of the summer is

$$\frac{d}{dt}y(t) = x(t) + ay(t)$$

Writing this first-order differential equation in the standard form gives us the system differential equation

$$\frac{d}{dt}y(t) - ay(t) = x(t).$$

Example 2.17

Consider the block diagram shown in Figure 2.30.

What is the system described by this block diagram?

FIGURE 2.29   System for Example 2.16.

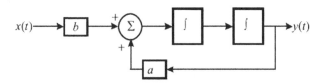

FIGURE 2.30   System for Example 2.17.

SOLUTION

The input to the first integrator is the second derivative of $y(t)$. At the summer

$$\frac{d^2}{dt}y(t) = bx(t) + ay(t)$$

or

$$\frac{d^2}{dt}y(t) - ay(t) = bx(t),$$

which is the differential equation described by the block diagram given in Figure 2.30.

2.18 FROM DIFFERENTIAL EQUATIONS TO BLOCK DIAGRAMS

This method is also best understood by using examples.

Example 2.18

Consider the following differential equation

$$\frac{d}{dt}y(t) - ay(t) = x(t).$$

What is the block diagram that represents the system or the differential equation?

SOLUTION

Let D represent $\frac{d}{dt}$, D^2 represent $\frac{d^2}{dt}$, and so on.

Therefore, if we substitute in the differential equation given we get

$$Dy(t) - ay(t) = x(t).$$

Multiply by D^{-1}, where 1 is the highest power in the above differential equation, the above equation becomes

$$y(t) - aD^{-1}y(t) = D^{-1}x(t).$$

We then solve for $y(t)$ to get

$$y(t) = D^{-1}\left[x(t) + ay(t)\right].$$

The differential equation we started with has only a first derivative, and in the block diagram we will need one integrator, one input $x(t)$, and one output $y(t)$. We will also have a summer at each input of any integrator used in the block diagram and use that as a rule for constructing block diagrams.

We will take the following steps to draw the block diagram.

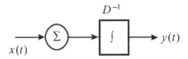

FIGURE 2.31   System for Example 2.18.

STEP 1

This step is represented graphically, according to the rule just explained, shown in Figure 2.31. The "D" letter on the integrator block in the figure is just to aid in understanding the procedure.

STEP 2

The block shown in Figure 2.32 is a direct implementation of the equation

$$y(t) = D^{-1}[x(t) + ay(t)].$$

Notice that the terms $x(t)$ and $ay(t)$ are inputs to the summer that comes before the integrator.

Example 2.19

Consider the following differential equation

$$\frac{d^2}{dt}y(t) - ay(t) = bx(t).$$

What is the block diagram that represents this system?

SOLUTION

Using the same procedure as explained before, we can write the differential equation as

$$D^2y(t) - ay(t) = bx(t).$$

Multiply the above equation by D^{-2}, where 2 is the highest power in the equation above, to get

$$y(t) - aD^{-2}y(t) = bD^{-2}x(t).$$

FIGURE 2.32   System for Example 2.18.

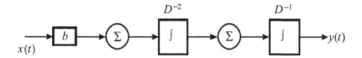

FIGURE 2.33   System for Example 2.19.

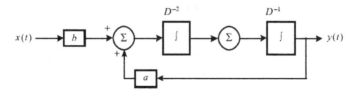

FIGURE 2.34   System for Example 2.19.

We then solve for y(t) to get

$$y(t) = D^{-2}\left[ay(t) + bx(t)\right].$$

Since we were given a second-order differential equation, we will have two integrators in the block diagram. We will also have the input x(t) and the output y(t). Each integrator will have a summer at its input.

STEP 1
This step is accomplished by drawing the initial block shown in Figure 2.33.

STEP 2
The terms bx(t) and ay(t) are associated with D^{-2} and are inputs to the summer, which is the input of the first integrator from the left.

We do not see a D^{-1} term in the above equation and therefore the term associated with D^{-1} is zero. Zero means nothing is supplied as input to the second integrator other than the straight path from the first integrator. The final block representation is shown in Figure 2.34.

2.19   THE IMPULSE RESPONSE

To find the response to an input x(t), the output of the system y(t) will have the following form

$$y(t) = \int_{-\infty}^{+\infty} x(\tau)h(t-\tau)d\tau,$$

where h(t) is the impulse response of the system when x(t) = δ(t). Therefore, we can say that if x(t) = δ(t), then y(t) = h(t), the impulse response.

The following examples will show us how to find h(t).

Example 2.20

Consider the first-order differential equation

$$\frac{d}{dt}y(t) + 4y(t) = x(t)$$

Find the impulse response for this system.

SOLUTION

To find the impulse response for the given system, we set the input $x(t)$ equal to $\delta(t)$ and find $y(t)$, which is then $h(t)$, the impulse response. Therefore, the given equation representing the system in this case can be written as

$$\frac{d}{dt}(h(t)) + 4h(t) = \delta(t).$$

Again, the total solution for $h(t)$, as we have seen before, will have two parts: the homogenous part and the particular part.

For the homogenous solution we set $x(t)$ to zero and this kills the $\delta(t)$ signal. Therefore, we have

$$\frac{d}{dt}(h(t)) + 4h(t) = 0.$$

The auxiliary (characteristic) equation is

$$m + 4 = 0,$$

which indicates that $m = -4$ and the homogenous solution is

$$h_h(t) = c_1 e^{-4t} \quad t > 0.$$

To find c_1 we need the total solution for $h(t)$ first.

For the particular solution we predict that the solution has the form

$$h_p(t) = 0$$

because this is always the case in most real systems, where the order of the derivative in the output $y(t)$ is greater than the order of the derivative in the input $x(t)$.

Therefore, the total solution is

$$h(t) = h_h(t) + h_p(t)$$

$$h(t) = c_1 e^{-4t} \quad t > 0,$$

or we can write instead

$$h(t) = c_1 e^{-4t} u(t),$$

where $u(t)$ is the unit step signal that when multiplied by or other signal will force the other signal to start at $t = 0$ without any other alteration in the other signal.

Finally we have to find c_1. To do so we substitute the total solution for $h(t)$ in the original differential equation noting that the derivative of $u(t)$ is $\delta(t)$. The derivative of $h(t)$ is

$$\frac{d}{dt}h(t) = c_1(-4)e^{-4t}u(t) + c_1e^{-4t}\delta(t),$$

where we used the product rule for derivatives.

Therefore, substituting back in the differential equation gives

$$\left[c_1(-4)e^{-4t}u(t) + c_1e^{-4t}\delta(t)\right] + 4\left[c_1e^{-4t}u(t)\right] = \delta(t).$$

Remember that the $\delta(t)$ signal is only valid at $t = 0$, and in this case any function multiplied by $\delta(t)$ is evaluated at $t = 0$. Therefore

$$\left[c_1(-4)e^{-4t}u(t) + c_1\delta(t)\right] + 4\left[c_1e^{-4t}u(t)\right] = \delta(t).$$

The above equation reduces to

$$c_1\delta(t) = \delta(t).$$

Equating coefficients we find that $c_1 = 1$, and the final solution is

$$h(t) = y(t) = e^{-4t} \; t > 0.$$

2.20   SOME INSIGHTS: CALCULATING $y(t)$

If we study carefully the solution of the system in Example 2.20, we see that the output $y(t)$ is

$$y(t) = e^{-4t} \; t > 0.$$

As time approaches infinity, the output will approach the value zero. In this sense, we say the output is stable for our particular input. For first-order systems (that is described by first-order differential equations), the output will have one exponential term at the most. This exponential term is called the transient of the system. For second-order systems, the output will have two transients (two exponential terms) at the most. For third-order systems we will have three exponential terms, and so on.

In many systems of order greater than two, and for the purpose of analysis and design, we can reduce the order of the system at hand to a second-order system due to the fast decay of some transients. The solution for the output for these systems is in the following form

$$y(t) = c_1e^{\alpha_1 t} + c_2e^{\alpha_2 t}.$$

The stability of the system is determined by the values of α_1 and α_2. If any of the α's are positive, the output $y(t)$ will grow wild as t approaches infinity. So if the α's are all negative then the output $y(t)$ will decay gradually and stay at a fixed value as time progresses. The α's are called the eigenvalues of the system. Therefore, we can say that an LTI system is stable if the eigenvalues of the system are all negative.

2.20.1   How Can We Find These Eigenvalues?

An LTI system can always be represented by a linear differential equation with constant coefficients

$$a_n \frac{d^n}{dt} y(t) + a_{a-1} \frac{d^{n-1}}{dt} y(t) + a_{n-2} \frac{d^{n-2}}{dt} y(t) + \ldots + a_0 y(t)$$

$$= b_m \frac{d^m}{dt} x(t) + b_{m-1} \frac{d^{m-(1)}}{dt} x(t) + \ldots + b_0 x(t).$$

We can look at the auxiliary algebraic equation by setting the input $x(t)$ to zero

$$a_n \frac{d^n}{dt} y(t) + a_{n-1} \frac{d^{n-1}}{dt} y(t) + \frac{d^{n-2}}{dt} y(t) + \ldots + a_0 y(t) = 0$$

and letting $\frac{d}{dt} = m$ to get

$$a_n m^n y(t) + a_{n-1} m^{n-1} y(t) + \ldots + a_0 y(t) = 0.$$

We can factor out $y(t)$ as

$$y(t) \left[a_n m^n + a_{n-1} m^{n-1} + \ldots + a_0 \right] = 0.$$

$y(t)$ cannot be zero (in which case the output of the system would be zero at all times), and therefore,

$$\left[a_n m^n + a_{n-1} m^{n-1} + \ldots + a_0 \right] = 0$$

or

$$\left[m^n + \frac{a_{n-1}}{a_n} m^{n-1} + \ldots + \frac{a_0}{a_n} \right] = 0.$$

This is a nth-order algebraic equation with n roots. These are the eigenvalues of the system.

2.20.2   Stability and Eigenvalues

To summarize, any LTI system can be modeled by a linear differential equation with constant coefficients. The auxiliary algebraic equation that can be obtained from the differential equation will have a number of roots called the eigenvalues of the system. The stability of the system is determined by these roots. These roots may be real or complex. If *all* the real parts of the roots are negative, then the system is stable. If *any* of the real parts of the roots is positive, then the system is unstable. The eigenvalues of the system are responsible for the shape of the output y(t). They dictate the shape of the transients of the system as well.

2.21   END-OF-CHAPTER EXAMPLES

EOCE 2.1

Consider the following systems.

1. $y(t) = \dfrac{1}{x(t)}$

2. $y(t) = t\,\sin(x(t))$

3. $y(t) = t\,\ln(x(t))$

Are the above systems linear?

SOLUTION

For the first system

$$y_1(t) = \frac{1}{x_1(t)}$$

$$y_2(t) = \frac{1}{x_2(t)}$$

$$y(t) = \frac{1}{\alpha x_1(t) + \beta x_2(t)}$$

Is $y(t) = \alpha y_1(t) + \beta y_2(t)$?

$$\alpha y_1(t) = \frac{\alpha}{x_1(t)}$$

$$\beta y_2(t) = \frac{\beta}{x_2(t)}$$

$$y(t) = \frac{1}{\alpha x_1(t) + \beta x_2(t)} \neq \frac{\alpha}{x_1(t)} + \frac{\beta}{x_2(t)}$$

Therefore, the system is not linear.

For the second system

$$y_1(t) = t \, \sin\big(x_1(t)\big)$$

$$y_2(t) = t \, \sin\big(x_2(t)\big)$$

$$y(t) = t \, \sin\big(\alpha x_1(t) + \beta x_2(t)\big)$$

Is $y(t) = \alpha y_1(t) + \beta y_2(t)$?

$$\alpha y_1(t) = \alpha t \, \sin\big(x_1(t)\big)$$

$$\beta y_2(t) = \beta t \, \sin\big(x_2(t)\big)$$

$$y(t) = t \, \sin\big(\alpha x_1(t) + \beta x_2(t)\big) \neq \alpha t \, \sin\big(x_1(t)\big) + \beta t \, \sin\big(x_2(t)\big)$$

Therefore, the system is not linear.
　For the third system

$$y_1(t) = t \, \ln\big(x_1(t)\big)$$

$$y_2(t) = t \, \ln\big(x_2(t)\big)$$

$$y(t) = t \, \ln\big(\alpha x_1(t) + \beta x_2(t)\big)$$

Is $y(t) = \alpha y_1(t) + \beta y_2(t)$?

$$\alpha y_1(t) = \alpha t \, \ln\big(x_1(t)\big)$$

$$\beta y_2(t) = \beta t \, \ln\big(x_2(t)\big)$$

$$y(t) = t \, \ln\big(\alpha x_1(t) + \beta x_2(t)\big) \neq \alpha t \, \ln\big(x_1(t)\big) + \beta t \, \ln\big(x_2(t)\big)$$

Therefore, the system is not linear.

EOCE 2.2

Consider EOCE 2.1 again. Are the systems time-invariant?

SOLUTION

For the first system

$$y_1(t) = \frac{1}{x_1(t)}$$

$$y_1(t - t_0) = \frac{1}{x_1(t - t_0)}$$

$$y_2(t) = \frac{1}{x_1(t-t_0)}$$

$$y_1(t-t_0) = \frac{1}{x_1(t-t_0)} = y_2(t) = \frac{1}{x_1(t-t_0)}$$

Therefore, the system is time-invariant.
For the second system

$$y_1(t) = t \sin(x_1(t))$$

$$y_1(t-t_0) = (t-t_0) \sin(x_1(t-t_0))$$

$$y_2(t) = t \sin(x_1(t-t_0))$$

$$y_1(t-t_0) = (t-t_0) \sin(x_1(t-t_0)) \neq y_2(t) = t \sin(x_1(t-t_0))$$

Therefore, the system is not time-invariant.
For the third system

$$y_1(t) = t \ln(x_1(t))$$

$$y_1(t-t_0) = (t-t_0) \ln(x_1(t-t_0))$$

$$y_2(t) = t \ln(x_1(t-t_0))$$

$$y_1(t-t_0) = (t-t_0) \ln(x_1(t-t_0)) \neq y_2(t) = t \ln(x_1(t-t_0))$$

Therefore, the system is not time-invariant.

EOCE 2.3

Consider the systems represented in block diagrams shown in Figure 2.35.
Find the output $y(t)$ using convolution for both systems.

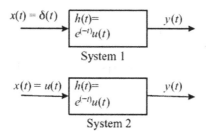

FIGURE 2.35 Systems for EOCE 2.3.

SOLUTION

SYSTEM 1

We can use the convolution integral to find the output $y(t)$ as

$$y(t) = \int_{-\infty}^{+\infty} x(\tau)h(t-\tau)d\tau = \int_{-\infty}^{+\infty} h(\tau)x(t-\tau)d\tau$$

$$y(t) = \int_{-\infty}^{+\infty} e^{-\tau}u(\tau)\delta(t-\tau)d\tau.$$

Using the sifting property

$$\int_{-\infty}^{+\infty} x(\tau)\delta(t-\tau)d\tau = x(t)$$

and noticing that the impulse is applied at $\tau = t$, the output becomes

$$y(t) = e^{-t}u(t).$$

SYSTEM 2

In this case, we also use the convolution integral and write

$$y(t) = \int_{-\infty}^{+\infty} h(\tau)x(t-\tau)d\tau = \int_{-\infty}^{+\infty} e^{-\tau}u(\tau)u(t-\tau)d\tau.$$

Notice that $u(t-\tau)$ is a unit step function that is reflected and shifted by t units. Also $h(\tau)$ is stationary with $u(t-\tau)$ moving to the right. Overlapping will start at $\tau = 0$ and stop at some $\tau = t$.

Remember also that $u(t-\tau)$ and $u(\tau)$ are multiplied under the integrand. Therefore,

$$y(t)\int_{0}^{t} e^{-\tau}d\tau = -\left[e^{-\tau}\right]_{0}^{t}$$

$$y(t) = -\left[e^{-t} - 1\right] = \left[1 - e^{-t}\right] t \ge 0.$$

EOCE 2.4

Consider the block diagram shown in Figure 2.36.
Find the output $y(t)$.

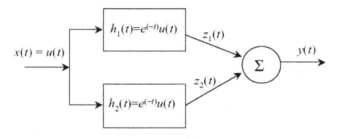

FIGURE 2.36 System for EOCE 2.4.

SOLUTION

This is the case of a parallel system and the output y(t) is calculated as

$$y(t) = z_1(t) + z_2(t)$$

with

$$z_1(t) = x(t) * h_1(t)$$

and

$$z_2(t) = x(t) * h_2(t).$$

Using the convolution integral we write for $z_1(t)$

$$z_1(t) = \int_{-\infty}^{+\infty} h_1(\tau) x(t-\tau) d\tau = \int_{-\infty}^{+\infty} e^{-\tau} u(\tau) u(t-\tau) d\tau$$

$$z_1(t) = \int_0^t e^{-\tau} d\tau = -\left[e^{-\tau} \right]_0^t,$$

and finally

$$z_1(t) = \left[1 - e^{-\tau} \right] t \ge 0.$$

For $z_2(t)$ we write

$$z_2(t) = \int_{-\infty}^{+\infty} h_2(\tau) x(t-\tau) d\tau = \int_{-\infty}^{+\infty} e^{-\tau} u(\tau) u(t-\tau) d\tau$$

$$z_2(t) = \int_0^t e^{-\tau} d\tau = -\left[e^{-\tau} \right]_0^t,$$

and

$$z_2(t) = \left[1 - e^{-\tau}\right] t \geq 0.$$

The output is then

$$y(t) = z_1(t) + z_2(t)$$

$$y(t) = \left[1 - e^{-t}\right] + \left[1 - e^{-t}\right] = \left[2 - 2e^{-t}\right] t \geq 0.$$

Using MATLAB script EOCE2_4, we can plot $h_1(t)$, $h_2(t)$, $z_2(t)$, and $y(t)$ for the impulse input $x(t)$
    The plots are shown in Figure 2.37.

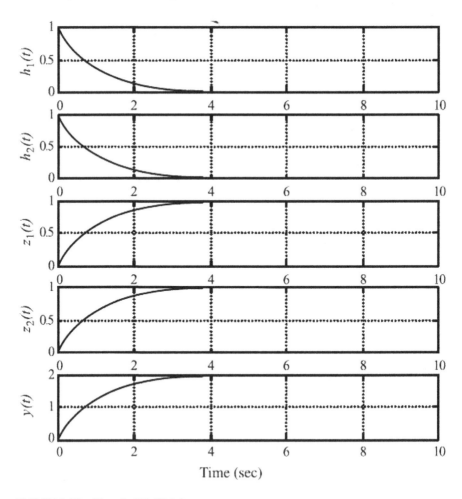

FIGURE 2.37 Plots for EOCE 2.4.

EOCE 2.5

Consider the block diagram shown in Figure 2.38.

Use $h_1(t)$ and $h_2(t)$ as given in EOCE 2.4 with $x(t)$ as a unit step input and find the output $y(t)$.

SOLUTION

We start with the intermediate output $z(t)$ and write

$$z(t) = \int_{-\infty}^{+\infty} h_1(\tau) x(t-\tau) d\tau = \int_{-\infty}^{+\infty} e^{-\tau} u(\tau) u(t-\tau) d\tau$$

$$z(t) = \int_0^t e^{-\tau} d\tau = -\left[e^{-\tau}\right]_0^t$$

and finally

$$z(t) = \left[1 - e^{-t}\right] t \geq 0.$$

But $y(t)$ is the convolution between $z(t)$ and $h_2(t)$

$$y(t) = z(t) * h_2(t)$$

$$y(t) = \int_{-\infty}^{+\infty} h_2(\tau) z(t-\tau) d\tau = \int_{-\infty}^{+\infty} \left(e^{-\tau}\right) u(\tau)\left(1 - e^{-(t-\tau)}\right) u(t-\tau) d\tau,$$

which simplifies to

$$y(t) = \int_0^t \left(e^{-\tau}\right)\left(1 - e^{-(t-\tau)}\right) d\tau = \int_0^t \left(e^{-\tau} - e^{-t}\right) d\tau$$

$$y(t) = \left[1 - e^{-t}\right] - \left[e^{-t}\tau\right]_0^t.$$

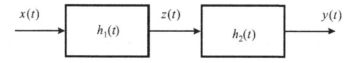

FIGURE 2.38   System for EOCE 2.5.

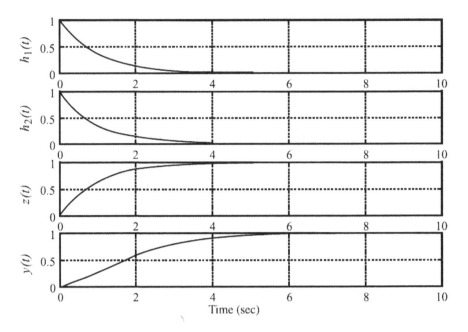

FIGURE 2.39 Plots for EOCE 2.5.

Finally,

$$y(t) = \left[1 - e^{-t}\right] - \left[e^{-t}t\right] = \left[1 - e^{-t}(1 + t)\right] t \geq 0.$$

For this impulse input, we can use the MATLAB script EOCE2_5 to plot $z(t)$ and $y(t)$.

The plots are shown in Figure 2.39.

EOCE 2.6

Find the graphical convolution

$$y(t) = x_1(t) * x_2(t)$$

where the signals $x_1(t)$ and $x_2(t)$ are shown in Figure 2.40.

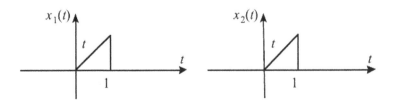

FIGURE 2.40 Signals for EOCE 2.6.

SOLUTION

We start by mathematically writing the two signals $x_1(t)$ and $x_2(t)$ as

$$x_1(t) = x_2(t) = \begin{cases} t & 0 < t < 1 \\ 0 & \text{otherwise} \end{cases}$$

with

$$x_1(\tau) = \begin{cases} \tau & 0 < \tau < 1 \\ 0 & \text{otherwise} \end{cases}.$$

We will fix $x_1(t)$ and slide $x_2(t)$. Therefore,

$$x_2(t - \tau) = \begin{cases} t - \tau & 0 < t - \tau < 1 \\ 0 & \text{otherwise} \end{cases}$$

or

$$x_2(t - \tau) = \begin{cases} t - \tau & t - 1 < \tau < 1 \\ 0 & \text{otherwise} \end{cases}.$$

For $t < 2$, there is no overlap and

$$y(t) = 0.$$

For $0 \le t < 1$, the two graphs are shown in Figure 2.41 where

$$y(t) = \int_0^t \tau(t - \tau) d\tau$$

$$y(t) = \left(\frac{t\tau^2}{2} - \frac{\tau^3}{3} \right) \Big|_0^t ,$$

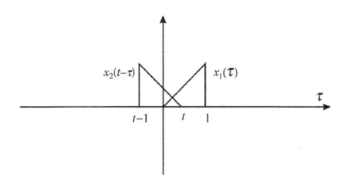

FIGURE 2.41   Signals for EOCE 2.6.

which results in

$$y(t) = \frac{t^3}{6}.$$

The plot of $y(t)$ for $0 < t < 1$, using the MATLAB script EOCE2_61, is shown in Figure 2.42.

For $1 \leq t < 2$, the graphs for this range are now shown in Figure 2.43. The overlapping region is between $\tau = t - 1$ and $\tau = 1$. The convolution results in

$$y(t) = \int_{t-1}^{1} \tau(t - \tau) d\tau$$

$$y(t) = \left(\frac{t\tau^2}{2} - \frac{\tau^3}{3} \right)_{t-1}^{1},$$

FIGURE 2.42   Plot for EOCE 2.6.

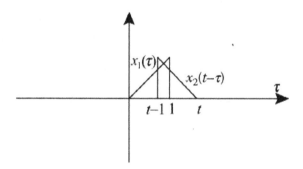

FIGURE 2.43   Signal for EOCE 2.6.

which results in

$$y(t) = \frac{-5t^3 + 12t^2 - 6t}{6}.$$

Let us plot $y(t)$ for $1 < t < 2$ using the MATLAB script EOCE2_62.
The plot is shown in Figure 2.44.
For $t > 2$, $y(t) = 0$.

FIGURE 2.44   Plot for EOCE 2.6.

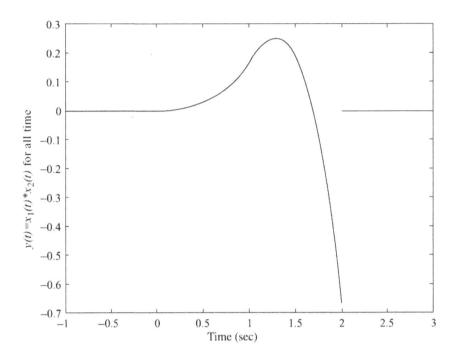

FIGURE 2.45 Plot for EOCE 2.6.

Now we can put all pieces of y(t) together using the MATLAB script EOCE2_63. The plots are shown in Figure 2.45.

EOCE 2.7

Find the impulse response for the following systems

1. $\dfrac{d}{dt} y(t) + 16y(t) = x(t)$

2. $\dfrac{d^2}{dt} y(t) + 8 \dfrac{d}{dt} y(t) + 16y(t) = x(t)$

SOLUTION

Let us consider the first system. With an impulsive input, the output of the system y(t) is the impulse response h(t). h(t) has two parts: the homogenous part and the particular part. For most real physical systems, the order of the derivative involving x(t), the input, is less than the order of the derivative of the output y(t). This is the case here, and hence the particular solution for h(t) is zero. The homogenous part of the given system is

$$\frac{d}{dt} y(t) + 16y(t) = 0$$

The characteristic equation is

$$m + 16 = 0,$$

for which m is -16.

Therefore, the homogenous part of $h(t)$ is

$$h_h(t) = c_1 e^{(-16t)} u(t).$$

The total solution of $h(t)$ is the sum of the homogenous part and the particular part

$$h(t) = h_h(t) + h_p(t)$$

$$h(t) = c_1 e^{(-16t)} u(t).$$

We need to calculate the constant c_1. To do that we substitute the solution $h(t)$ in the given system to get

$$\left[-16 c_1 e^{(-16t)} u(t) + c_1 e^{(-16t)} \delta(t) \right] + 16 \left[c_1 e^{(-16t)} \right] u(t) = \delta(t).$$

Note that the middle term at the left part of the equal sign is multiplied by the impulse function that is applied at $t = 0$. At this point we use the relation $f(t)\delta(t) = f(0)\delta(t)$, where $f(t)$ is any function, to simplify the above equation. We will get

$$c_1 \delta(t) = \delta(t).$$

Comparing coefficients, we clearly see that $c_1 = 1$. Therefore

$$h(t) = e^{(-16t)} u(t).$$

This response can be seen graphically using the MATLAB script EOCE2_71.

The plot is shown in Figure 2.46.

For the second system, and using the same arguments as we did for the first system, $h_p(t)$ is zero. The homogenous equation is given as

$$\frac{d^2}{dt} y(t) + 8 \frac{d}{dt} y(t) + 16 y(t) = 0.$$

Remember that when you calculate the impulse response, $y(t)$ is $h(t)$. The characteristic equation is

$$m^2 + 8m + 16 = 0,$$

with the double repeated roots at $m = -4$.

The homogenous solution in this case has the form

$$h_h(t) = c_1 e^{(-4t)} u(t) + c_2 t e^{(-4t)} u(t).$$

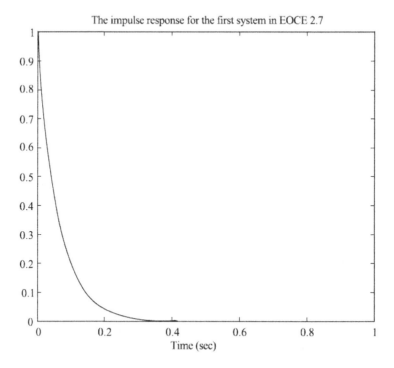

FIGURE 2.46   Signal for EOCE 2.7.

We next substitute in the original system to get

$$\left\{\left[16c_1e^{-4t}u(t) - 4c_1e^{(-4t)}\delta(t)\right] + \left[-4c_1e^{(-4t)}\delta(t) + c_1e^{(-4t)}\frac{d}{dt}\delta(t)\right]\right.$$

$$+\left[-4c_2e^{(-4t)}u(t) + c_2e^{(-4t)}\delta(t)\right] + \left[16c_2te^{(-4t)}u(t) - 4c_2te^{(-4t)}\delta(t)\right]$$

$$+\left.\left[-4c_2e^{(-4t)}u(t) + c_2e^{(-4t)}\delta(t)c_2te^{(-4t)}\frac{d}{dt}\delta(t) - 4c_2te^{(-4t)}\delta(t)\right]\right\}$$

$$+\left\{-32c_1e^{(-4t)}u(t) + 8c_1e^{(-4t)}\delta(t) + 8c_2e^{(-4t)}u(t) - 32c_2te^{(-4t)}u(t)\right.$$

$$+8c_2te^{(-4t)}\delta(t)\Big\} + \left\{16c_1e^{(-4t)}u(t) + 16c_2te^{(-4t)}u(t)\right\} = \delta(t).$$

We can clean the above to get

$$-4c_1e^{(-4t)}\delta(t) - 4c_1e^{(-4t)}\delta(t) + c_1e^{(-4t)}\frac{d}{dt}\delta(t) - c_2e^{(-4t)}\delta(t) - 4c_1te^{(-4t)}\delta(t)$$

$$+c_2e^{(-4t)}\delta(t) + c_2te^{(-4t)}\frac{d}{dt}\delta(t) - 4c_2te^{(-4t)}\delta(t) + 8c_1e^{(-4t)}\delta(t)$$

$$+8c_2te^{(-4t)}\delta(t) = \delta(t).$$

With $f(t)\delta(t) = f(0)\delta(t)$, where $f(t)$ is any function, we can simplify further and write

$$-4c_1\delta(t) - 4c_1\delta(t) + c_1\frac{d}{dt}\delta(t) - c_2\delta(t) + c_2\delta(t) + 8c_1\delta(t) = \delta(t).$$

By equating coefficients for $\delta(t)$ we get

$$-4c_1 - 4c_1 + c_2 + c_2 + 8c_1 = 1,$$

and by equating coefficients for $\frac{d}{dt}\delta(t)$ we get

$$c_1 = 0.$$

Therefore, $c_2 = 1/2$ and the impulse response is

$$h(t) = (1/2)te^{(-4t)}u(t).$$

Using MATLAB we can plot this response as in Figure 2.47 by using the EOCE2_72 script.

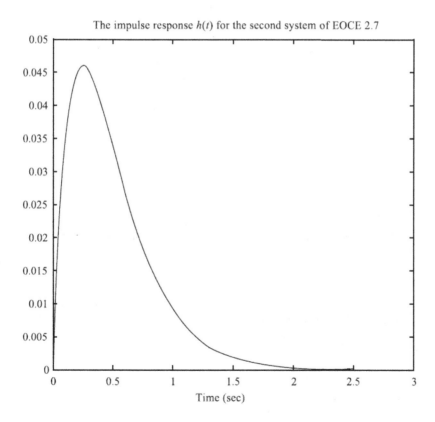

FIGURE 2.47   Signal for EOCE 2.7.

EOCE 2.8

Consider the following systems.

1. $\dfrac{d^2}{dt}y(t) + 8\dfrac{d}{dt}y(t) + 16y(t) = x(t)$

2. $\dfrac{d^2}{dt}y(t) + \dfrac{d}{dt}y(t) + y(t) = x(t)$

3. $\dfrac{d^2}{dt}y(t) + 5\dfrac{d}{dt}y(t) + 6y(t) = x(t)$

4. $\dfrac{d^2}{dt}y(t) + 16y(t) = x(t)$

5. $\dfrac{d^2}{dt}y(t) + 16\dfrac{d}{dt}y(t) = x(t)$

1. Discuss stability for each of the systems above.
2. What is the general form of the solution for the homogenous part of each system?

SOLUTION

SYSTEM 1

The characteristic equation is

$$m^2 + 8m + 16 = 0.$$

Using MATLAB we can find the roots by typing at the prompt

```
eigenvalues_for_system_1 = roots([1, 8, 16])
```

to get

```
eigenvalues_for_system_1 =
-4
-4
```

Since the eigenvalues are all negative, the system is stable. The general form of the solution is

$$y(t) = c_1 e^{(-4t)} + tc_2 e^{(-4t)} \text{ for } t \geq 0.$$

As you can see the solution, $y(t)$, converges to zero as $t \to \infty$. We can use MATLAB and type the following at the prompt

```
system_1_solution =dsolve ('D2y+8*Dy+16*y=0', 't')
```

to get

```
system_1_solution =
C1*exp (-4*t) + C2*exp (-4*t)*t
```

This solution agrees with the analytical solution presented previously.

SYSTEM 2

The characteristic equation is

$$m^2 + m + 1 = 0.$$

Using MATLAB we can find the roots as

```
eigenvalues_for_system_2 = roots([1, 1, 1])
```

to get

```
eigenvalues_for_system_2 =
-0.5000 + 0.8660i
-0.5000 - 0.8660i
```

Since the real parts of the eigenvalues are all negative, the system is stable. The general form of the solution is

$$y(t) = c_1 e^{[(-0.5+0.866j)t]} + c_2 e^{[(-0.5-0.866j)t]} \text{ for } t \geq 0$$

$$y(t) = c_1 e^{(-0.5t)}\left[\cos(0.866t) + j\,\sin(0.866t)\right]$$
$$+ c_2 e^{(-0.5)t}\left[\cos(0.866)t - j\,\sin(0.866t)\right] \text{ for } t \geq 0.$$

$y(t)$ converges to zero as $t \to \infty$.
We can use MATLAB again to solve system 2 and type at the prompt

```
system_2_solution=dsolve('D2y+Dy+y=0', 't')
```

to get

```
system_2_solution =
C1*exp(-1/2*t)*co(1/2*3^1/2)*t)+C2*exp(-1/2*t)
*sin(1/2*3^1/2)*t)
```

The two solutions are the same. Prove that.

SYSTEM 3

The characteristic equation is

$$m^2 + 5m + 6 = 0.$$

Using MATLAB we can find the roots as

```
eigenvalues_for_system_3 = roots([1, 5, 6])
```

to get

```
eigenvalues_for_system_3 =
 -3
 -2
```

Since all the roots are negative, the system is stable. The general form of the solution is

$$y(t) = c_1 e^{(-3t)} + c_2 e^{(-2t)} \text{ for } t \geq 0.$$

$y(t)$ converges to zero as $t \to \infty$. This means stability.
    With MATLAB we type at the prompt

```
system_3_solution=dsolve('D2y+5*Dy+6*y=0', 't')
```

to get

```
system_3_solution =
C1*exp(- 3*t)+C2*exp(- 2*t)
```

The two solutions agree.

SYSTEM 4

The characteristic equation is

$$m^2 + 16 = 0.$$

Using MATLAB we can find the roots by typing the following at the prompt

```
eigenvalues_for_system_4 = roots([1, 0, 16])
```

to get

```
eigenvalues_for_system_4 =
0 + 4 .0 0 0 0 i
0 - 4.0000i
```

Since there are no real parts for the roots, and since the roots are all imaginary, the system response is pure sinusoidal. In this case, the system is on the verge of instability and we consider it unstable. The general form of the solution is

$$y(t) = c_1 e^{[(4j)t]} + c_2 e^{[(-4j)t]} \text{ for } t \geq 0$$

$$y(t) = c_1 \left[\cos(4t) + j\sin(4t)\right] + c_2 \left[\cos(-4t) + j\sin(-4t)\right].$$

$y(t)$ does not converge to zero as $t \to \infty$. Pure sinusoidal signals are bounded and they do not die as time increases.
    With MATLAB we type at the prompt

```
system_4_solution = dsolve ('D2y+16*y=0 ' , 't')
```

to get

```
system_4_solution =
C1*sin(4*t)+C2*cos(4*t)
```

Try to prove that the two solutions are the same.

SYSTEM 5

The characteristic equation is

$$m^2 + 16m = 0.$$

Using MATLAB we can find the roots by typing the following at the MATLAB prompt.

```
eigenvalues_for_system_5 = roots([1, 16, 0])
```

to get

```
eigenvalues_for_system_5 =
0
-16
```

Since one of the roots is at zero and the other is negative (no real roots in the positive side of the complex plane), the system is on the verge of stability. The general form of the solution is

$$y(t) = c_1 e^{(0t)} + c_2 e^{(-16t)} \quad \text{for } t \ge 0$$

$$y(t) = c_1 + c_2 e^{(-16t)} \quad \text{for } t \ge 0.$$

As $t \to \infty$, $y(t)$ approaches $c_1 + c_2$, a constant. If the first root is one step in the right-half plane of the complex plane, the system would be unstable.

With MATLAB we type at the prompt

```
system_5_solution=dsolve('D2y+16*Dy=0', 't')
```

to get

```
system_5_solution =
C1 + C2*exp(-16*t)
```

That is what we have analytically.

EOCE 2.9

Consider the following system described by the differential equation

$$\frac{d^2}{dt} y(t) + \frac{d}{dt} y(t) = 2t$$

with the initial conditions

$$y(0) = 0 \quad \text{and} \quad \frac{d}{dt}y(0) = 0.$$

What is the solution $y(t)$?

SOLUTION

The solution was found earlier in the chapter as

$$y(t) = y_h(t) + y_p(t) = -2e^{-t} + 2 - 2t + t^2.$$

We can type at the MATLAB prompt the following command

```
y_total = dsolve('D2y+Dy=2*t','y(0)=0','Dy(0) = 0')
```

to get

```
y_total =
(-2*t*exp(t)+t^2*exp(t)+2*exp(t)-2)/exp(t)
```

which agrees with the analytical solution presented above and explained in Example 2.14.

EOCE 2.10

Consider the following system described by the following differential equation

$$\frac{d^3}{dt}y(t) - y(t) = \sin t$$

with initial conditions

$$y(0) = 0, \quad \frac{d}{dt}y(0) = 0, \quad \text{and} \quad \frac{d^2}{dt}y(0) = 0.$$

What is the output of the system?

SOLUTION

Again the solution was derived in Example 2.15:

$$y(t) = 0.159e^t + e^{-0.5t}\left[-0.659\cos(0.866t) + 0.013\sin(0.866t)\right]$$

$$+1/2\cos t - 1/2\sin t.$$

We can use MATLAB and type at the prompt

```
y_total=dsolve ('D3y-y=sin(t)','y (0)=0',…
'Dy(0)=0','D2y (0)=0');
y_total=simple (y_total)% to simplify the result
```

to get

```
y_total =
1/6*exp(t)-1/2*sin(t)+1/2*cos(t)-2/3*exp
(-1/2*t)*cos(1/2*3^(1/2)*t)
```

The two solutions seem to be different as we look at them. Is that true? One way to know is to plot the two solutions using the following MATLAB script EOCE2_10. The plots are shown in Figure 2.48.

The error you see in Figure 2.48 is due to truncation in the calculations.

You may also notice that the output is unstable. This is due to the presence of a positive root (a pole) at 1 as seen in the first term of the solution.

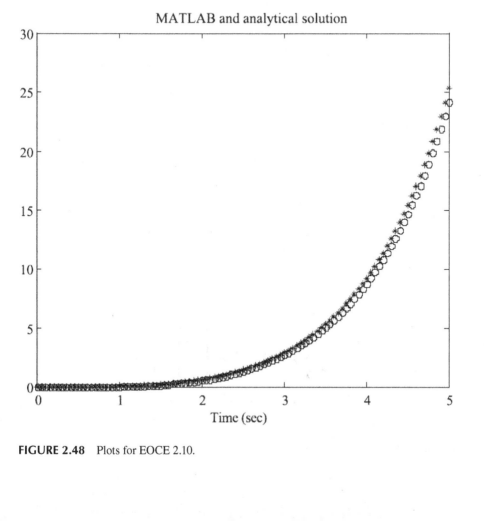

FIGURE 2.48 Plots for EOCE 2.10.

EOCE 2.11

Consider the system given in block diagram representation shown in Figure 2.49.

1. For what value(s) of k is the system stable?
2. Pick a value for k where the system is stable and find the impulse and the step responses.

SOLUTION

1. At the output of the first integrator the signal is

$$y_1(t) = \int \left[x(t) - ky_1(t) - y(t) \right] dt$$

where $y_1(t)$ is the output of the first integrator. In this case $y_1(t)$ is the derivative of $y(t)$.
   At the output of the second integrator

$$y(t) = \int y_1(t) \, dt.$$

Differentiating the first equation we get

$$\frac{d}{dt} y_1(t) = x(t) - ky_1(t) - y(t).$$

Differentiating the second equation twice we get

$$\frac{d^2}{dt} y(t) = \frac{d}{dt} y_1(t)$$

or

$$\frac{d^2}{dt} y(t) = x(t) - ky_1(t) - y(t) = -k \frac{d}{dt} y(t) + x(t) - y(t).$$

Finally

$$\frac{d^2}{dt} y(t) + k \frac{d}{dt} y(t) + y(t) = x(t).$$

FIGURE 2.49   System for EOCE 2.11.

If we apply the Routh test to this differential equation, we will have the following Routh array

$$
\begin{array}{cc}
1 & 1 \\
k & 0 \\
1 &
\end{array}
$$

which indicates that the system is stable for all positive values of k.

2. Using the MATLAB script EOCE2_11, we can find the step and the impulse responses for the system.

A sample run of the script is

```
enter a value for k, a negative number to exit: 1
```

and the plot is shown in Figure 2.50.

```
enter a value for k, a negative number to exit: 10
```

and the plot is in shown Figure 2.51.

```
enter a value for k, a negative number to exit: 100
```

and the plot is in shown Figure 2.52.

```
enter a value for k, a negative number to exit: -1
```

We can see that as k gets larger the system becomes slow in responding to the input signals. Notice also that for $k = 1$ the system is underdamped (has oscillations), and as k becomes larger it becomes undamped (no oscillations).

FIGURE 2.50 Plot for EOCE 2.11.

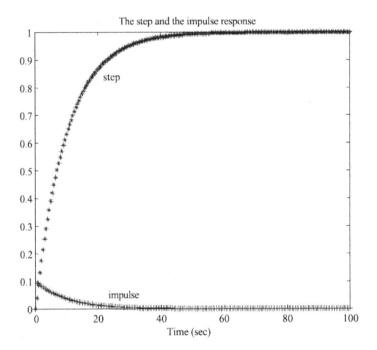

FIGURE 2.51   Plot for EOCE 2.11.

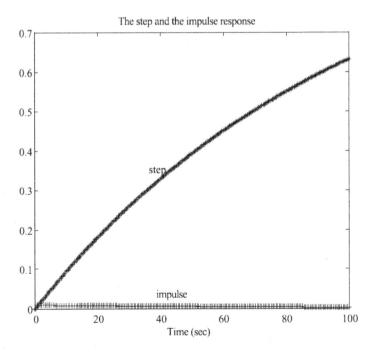

FIGURE 2.52   Plot for EOCE 2.11.

2.22 END-OF-CHAPTER PROBLEMS

EOCP 2.1

Consider the following systems. State whether each is linear or nonlinear and give your reasons. Also check if each is time-variant and give reasons.

1. $y(t) = x(t)\sqrt{x(t)}$
2. $y(t) = x(t) + \sqrt{x(t)}$
3. $y(t) = tx(t) + \sqrt{x(t)}$
4. $y(t) = tx(t) + e^{x(t)}$
5. $y(t) = tx(2t) + \sin(t)$

EOCP 2.2

Use the definition of stability provided in Section 2.8 to determine if the following systems are stable or not. Assume $x(t)$, the input to the system, is bounded.

1. $y(t) = \cos(x(t))$
2. $y(t) = t\cos(x(t))$
3. $y(t) = e^t \cos(x(t))$
4. $y(t) = e^{-t} \cos(x(t))$
5. $y(t) = te^{-t}\cos(x(t))$
6. $y(t) = te^{x(t)}\cos(x(t))$

7. $y(t) = \int_{-\infty}^{t} 10x(\tau)d\tau$

8. $y(t) = \int_{-\infty}^{t} x(\tau)e^{-\tau}d\tau$

9. $y(t) = \int_{-\infty}^{t} x(\tau)e^{\tau}d\tau$

10. $y(t) = \int_{-\infty}^{t} x(\tau)\tau e^{\tau}d\tau$

EOCP 2.3

Different systems are given below with specific inputs and the impulse response $h(t)$. Find $y(t)$, the output, using the convolution integral.

1. $x(t) = \delta(t)$, $h(t) = (e^{-t} + e^{-2t})u(t)$
2. $x(t) = u(t)$, $h(t) = (e^{-t} + e^{-2t})u(t)$

3. $x(t) = e^{-t}u(t)$, $h(t) = e^{-t}u(t)$
4. $x(t) = e^{-t}u(t) + u(t)$, $h(t) = e^{-2t}u(t)$
5. $x(t) = e^{-t}u(t) + \delta(t)$, $h(t) = e^{-2t}u(t)$
6. $x(t) = e^{-t}u(t) + \delta(t) + u(t)$, $h(t) = e^{-2t}u(t)$

EOCP 2.4

Use the convolution integral to find $y(t)$ in each of the systems shown in Figures 2.53–2.57.

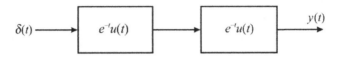

FIGURE 2.53   System for EOCP 2.4.

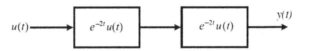

FIGURE 2.54   System for EOCP 2.4.

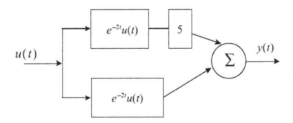

FIGURE 2.55   System for EOCP 2.4.

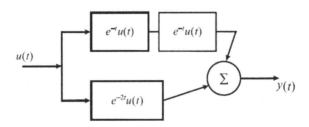

FIGURE 2.56   System for EOCP 2.4.

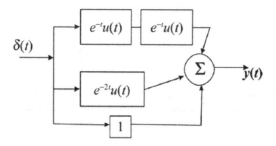

FIGURE 2.57 System for EOCP 2.4.

EOCP 2.5

Find the signal $y(t)$ that results from convolving $x(t)$ and $h(t)$ shown in Figures 2.58–2.64.

FIGURE 2.58 Signals for EOCP 2.5.

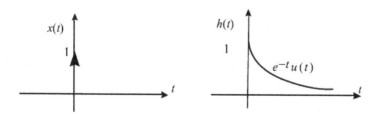

FIGURE 2.59 Signals for EOCP 2.5.

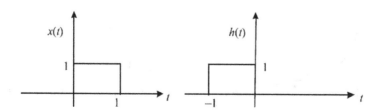

FIGURE 2.60 Signals for EOCP 2.5.

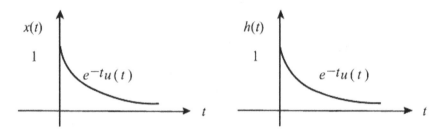

FIGURE 2.61   Signals for EOCP 2.5.

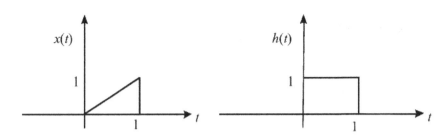

FIGURE 2.62   Signals for EOCP 2.5.

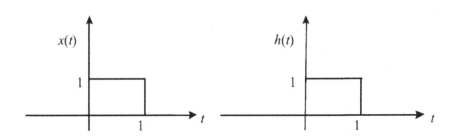

FIGURE 2.63   Signals for EOCP 2.5.

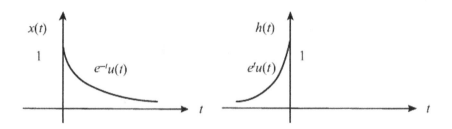

FIGURE 2.64   Signals for EOCP 2.5.

EOCP 2.6

Find the output $y(t)$ for the following systems with the indicated initial conditions.

1. $\dfrac{d}{dt}y(t)+\dfrac{1}{2}y(t)=0$ with $y(0)=-1$

2. $3\dfrac{d}{dt}y(t)-\dfrac{1}{2}y(t)=0$ with $y(0)=10$

3. $\dfrac{d^2}{dt}y(t)+\dfrac{1}{2}y(t)=0$ with $y(0)=-1$ and $\dfrac{d}{dt}y(0)=0$

4. $\dfrac{d^2}{dt}y(t)-\dfrac{1}{2}y(t)=0$ with $y(0)=1$ and $\dfrac{d}{dt}y(0)=-1$

5. $\dfrac{d^2}{dt}y(t)+\dfrac{1}{2}\dfrac{d}{dt}y(t)=0$ with $y(0)=-1$ and $\dfrac{d}{dt}y(0)=0$

6. $\dfrac{d^2}{dt}y(t)-\dfrac{1}{2}\dfrac{d}{dt}y(t)=0$ with $y(0)=-1$ and $y(0)=1$

7. $\dfrac{d^2}{dt}y(t)+\dfrac{1}{2}\dfrac{d}{dt}y(t)+y(t)=0$ with $y(0)=-1$ and $\dfrac{d}{dt}y(0)=0$

8. $\dfrac{d^2}{dt}y(t)+10\dfrac{d}{dt}y(t)+2y(t)=0$ with $y(0)=-1$ and $\dfrac{d}{dt}y(0)=-1$

9. $2\dfrac{d^2}{dt}y(t)-3\dfrac{d}{dt}y(t)=0$ with $y(0)=1$ and $\dfrac{d}{dt}y(0)=0$

10. $2\dfrac{d^2}{dt}y(t)+3\dfrac{d}{dt}y(t)-3y(t)=0$ with $y(0)=1$ and $\dfrac{d}{dt}y(0)=0$

EOCP 2.7

Use the roots of the characteristic equation to study the stability of the systems given in EOCP 2.6.

EOCP 2.8

Find the value(s) of k that makes each of the following systems stable.

1. $\dfrac{d}{dt}y(t)+\dfrac{1}{2}ky(t)=x(t)$

2. $3\dfrac{d}{dt}y(t)-\dfrac{1}{2}ky(t)=x(t)$

3. $\dfrac{d^2}{dt}y(t)+\dfrac{1}{2}ky(t)=x(t)$

4. $\dfrac{d^2}{dt}y(t)+\dfrac{1}{2}ky(t)=x(t)$

5. $\dfrac{d^2}{dt}y(t)+\dfrac{1}{2}(k+1)\dfrac{d}{dt}y(t)=x(t)$

6. $\dfrac{d^2}{dt}y(t)-\dfrac{1}{2}k\dfrac{d}{dt}y(t)=x(t)$

7. $\dfrac{d^2}{dt}y(t)+\dfrac{1}{2}\dfrac{d}{dt}y(t)+ky(t)=x(t)$

8. $\dfrac{d^2}{dt}y(t)+(k+1)\dfrac{d}{dt}y(t)+2y(t)=x(t)$

9. $2\dfrac{d^2}{dt}y(t)-3(k-1)\dfrac{d}{dt}y(t)=x(t)$

10. $2\dfrac{d^2}{dt}y(t)+(k+1)\dfrac{d}{dt}y(t)-3ky(t)=x(t)$

EOCP 2.9

Find the total solution of the following systems. Assume zero initial conditions.

1. $\dfrac{d^2y(t)}{dt}-\dfrac{1}{2}\dfrac{d}{dt}y(t)=\sin(t)$

2. $\dfrac{d^2}{dt}y(t)+\dfrac{1}{2}\dfrac{d}{dt}y(t)+y(t)=u(t)$

3. $\dfrac{d^2}{dt}y(t)+\dfrac{d}{dt}y(t)+2y(t)=e^{-t}u(t)$

4. $2\dfrac{d^2}{dt}y(t)-3\dfrac{d}{dt}y(t)=tu(t)$

EOCP 2.10

Find the impulse response for each of the following systems.

1. $\dfrac{d}{dt}y(t)+\dfrac{1}{2}y(t)=x(t)$

2. $3\dfrac{d}{dt}y(t)-\dfrac{1}{2}y(t)=x(t)$

3. $\dfrac{d^2}{dt}y(t)+\dfrac{1}{2}y(t)=x(t)$

4. $\dfrac{d^2}{dt}y(t)-\dfrac{1}{2}y(t)=x(t)$

5. $\dfrac{d^2}{dt}y(t)+\dfrac{1}{2}\dfrac{d}{dt}y(t)=x(t)$

EOCP 2.11

Draw the block diagrams for the following systems.

1. $\dfrac{d}{dt}y(t)+\dfrac{1}{2}y(t)=x(t)+\dfrac{d}{dt}x(t)$

2. $3\dfrac{d^2}{dt}y(t)-\dfrac{1}{2}\dfrac{d}{dt}y(t)=2x(t)$

3. $\dfrac{d^3}{dt}y(t)+y(t)=\dfrac{d^2}{dt}x(t)$

4. $\dfrac{d^3}{dt}y(t)-\dfrac{1}{2}\dfrac{d}{dt}y(t)=x(t)+\dfrac{d}{dt}x(t)$

5. $\dfrac{d^3}{dt}y(t)+\dfrac{1}{2}\dfrac{d^2}{dt}y(t)=x(t)$

EOCP 2.12

For the block diagrams shown in Figures 2.65–2.69, find the system that represents each.

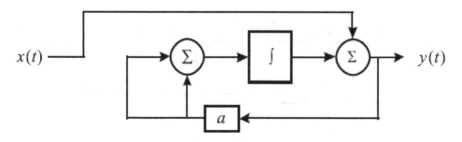

FIGURE 2.65 System for EOCP 2.12.

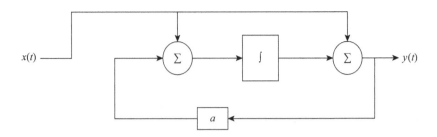

FIGURE 2.66 System for EOCP 2.12.

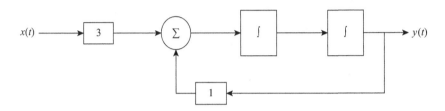

FIGURE 2.67 System for EOCP 2.12.

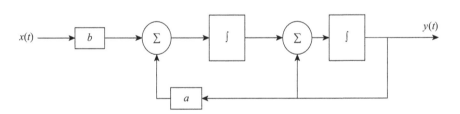

FIGURE 2.68 System for EOCP 2.12.

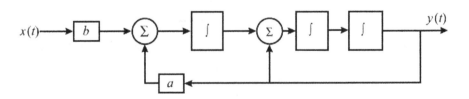

FIGURE 2.69 System for EOCP 2.12.

REFERENCES

Cogdell, J.R. *Foundations of Electrical Engineering*, 2nd ed., Englewood Cliffs, NJ: Prentice-Hall, 1996.

Denbigh, P. *System Analysis and Signal Processing*, Reading, MA: Addison-Wesley, 1998.

Golubitsky, M. and Dellnitz, M. *Linear Algebra and Differential Equations Using MATLAB*, Stamford, CT: Brooks/Cole, 1999.

Harman, T.L., Dabney, J., and Richert, N. *Advanced Engineering Mathematics with MATLAB*, Stamford, CT: Brooks/Cole, 2000.

Moore, H. *MATLAB for Engineers*, Upper Saddle River, NJ: Prentice Hall, 2007.

Nilson, W.J. and Riedel, S.A. *Electrical Circuits*, 6th ed., Englewood Cliffs, NJ: Prentice-Hall, 2000.

Perko, L. *Differential Equations and Dynamical Systems*, New York, NY: Springer-Verlag, 1991.

Phillips, C.L. and Parr, J.M. *Signals, Systems, and Transforms*, 2nd ed., Englewood Cliffs, NJ: Prentice-Hall, 1999.

Strum, R.D. and Kirk, D.E. *Contemporary Linear Systems*, Boston, MA: PWS, 1996.

The Math Works. *The Student Edition of MATLAB*, Englewood Cliffs, NJ: Prentice-Hall, 1997.

Woods, R.L. and Lawrence, K.L. *Modeling and Simulation of Dynamic Systems*, Englewood Cliffs, NJ: Prentice-Hall, 1997.

Wylie, R.C. and Barrett, C.L. *Advanced Engineering Mathematics*, 6th ed., New York, NY: McGraw-Hill, 1995.

Ziemer, R.E., Tranter, W.H., and Fannin, D.R. *Signals Systems Continuous and Discrete*, 4th ed., Englewood Cliffs, NJ: Prentice-Hall, 1998.

3 Fourier Series

3.1 REVIEW OF COMPLEX NUMBERS

This is a good point to give a brief review of complex numbers because this chapter and the chapters following are heavily involved with their arithmetic manipulation.

3.1.1 DEFINITION

By definition, the complex number j is the square root of -1. In general, the complex number $C = P + jQ$ consists of two parts: the real part P and the imaginary part Q. C can be represented in many forms as we will see later. If we are multiplying or dividing complex numbers we will prefer to use the polar form. If we are adding or subtracting complex numbers, we will rather use the rectangular form. The reason for that is the ease each form provides in the corresponding calculation.

Consider two complex numbers C_1 and C_2, where $C_1 = P_1 + jQ_1$ and $C_2 = P_2 + jQ_2$. These complex numbers are given in the rectangular form.

3.1.2 ADDITION

When we add two complex numbers we add their real parts and their imaginary parts separately to form the addition

$$C_1 + C_2 = (P_1 + P_2) + j(Q_1 + Q_2).$$

3.1.3 SUBTRACTION

When we subtract two complex numbers we subtract their real parts and their imaginary parts to form the subtraction

$$C_1 - C_2 = (P_1 - P_2) + j(Q_1 - Q_2).$$

3.1.4 MULTIPLICATION

Let us now consider the two complex numbers in polar form. In polar form C_1 and C_2 are represented as

$$C_1 = M_1 e^{j\theta_1}$$

and

$$C_2 = M_2 e^{j\theta_2},$$

119

where

$$M_1 = \sqrt{P_1^2 + Q_1^2}$$

$$M_2 = \sqrt{P_2^2 + Q_2^2}$$

and

$$\theta_1 = \tan^{-1}\left[\frac{Q_1}{P_1}\right]$$

$$\theta_2 = \tan^{-1}\left[\frac{Q_2}{P_2}\right].$$

We can also represent the complex numbers C_1 and C_2 as

$$M_1 \angle \theta_1$$

and

$$M_2 \angle \theta_2,$$

where M_1, M_2, θ_1, and θ_2 are as given above. The complex number $M_1 \angle \theta_1$ is read as a complex number with magnitude M_1 and phase angle θ_1. Complex numbers are easily multiplied in polar form as

$$C_1 C_2 = M_1 e^{j\theta_1} M_2 e^{j\theta_2}$$

$$C_1 C_2 = M_1 M_2 e^{j\theta_1} e^{j\theta_2} = M_1 M_2 e^{j(\theta_1 + \theta_2)}.$$

Or we can write

$$C_1 C_2 = M_1 M_2 \angle (\theta_1 + \theta_2).$$

If we have more than two complex numbers in polar form to be multiplied we use the same procedure. We multiply their magnitudes and add their phase angles to form the new product.

3.1.5 DIVISION

To divide two complex numbers we divide their magnitudes and subtract their phase angles.

$$C_1 / C_2 = [M_1 e^{j\theta_1}] / [M_2 e^{j\theta_2}]$$

$$C_1 / C_2 = [M_1 / M_2] e^{j(\theta_1 - \theta_2)}.$$

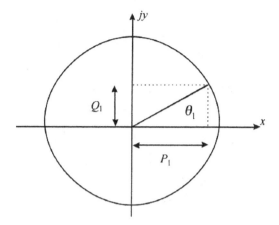

FIGURE 3.1 Rectangular form.

Or we can represent the division as

$$C_1/C_2 = [M_1/M_2] \angle (\theta_1 - \theta_2).$$

3.1.6 FROM RECTANGULAR TO POLAR

Consider the complex number

$$C_1 = P_1 + jQ_1.$$

This representation in the rectangular form is shown in Figure 3.1.

To convert to polar form we write C_1 as

$$C_1 = M_1 e^{j\theta_1} = M_1 \angle \theta_1$$

where

$$M_1 = \sqrt{P_1^2 + Q_1^2}$$

and

$$\theta_1 = \tan^{-1}\left[\frac{Q_1}{P_1}\right].$$

3.1.7 FROM POLAR TO RECTANGULAR

Consider the complex number

$$C_1 = M_1 e^{j\theta_1} = M_1 \angle \theta_1$$

in polar form as seen in Figure 3.2.

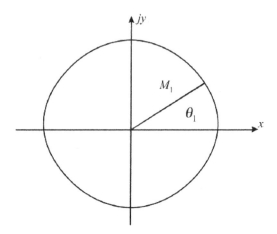

FIGURE 3.2   Polar form.

To convert to its equivalent rectangular form we write C_1 as

$$C_1 = P_1 + j\, Q_1,$$

where

$$P_1 = M_1 \mathrm{Cos}\theta_1$$

and

$$Q_1 = M_1 \mathrm{Sin}\theta_1.$$

These two equations can be solved simultaneously for P_1 and Q_1.

3.2   ORTHOGONAL FUNCTIONS

A set of signals, Φ_i, $i = 0, \pm 1, \pm 2, \ldots$ are orthogonal over an interval (a, b) if

$$\int_a^b \Phi_l \Phi_k^* dt = \begin{cases} E_k & l = k \\ 0 & l \neq k \end{cases} = E_k \delta(l - k) \tag{3.1}$$

where * indicates complex conjugate, and

$$\delta(l - k) = \begin{cases} 1 & l = k \\ 0 & l \neq k \end{cases}.$$

Example 3.1

Show that $\Phi_m = \sin(mt)$, $m = 1, 2, 3, \ldots$ form an orthogonal set on $(-\pi, \pi)$.

SOLUTION

We will use the defining integral for orthogonal functions to get

$$\int_{-\pi}^{\pi} \Phi_m(t)\Phi_n^*(t)\,dt = \int_{-\pi}^{\pi} \sin(mt)\sin(nt)\,dt$$

$$= 1/2\int_{-\pi}^{\pi} \cos(m-n)t\,dt - 1/2\int_{-\pi}^{\pi} \cos(m+n)t\,dt.$$

If $m = n$

$$\int_{-\pi}^{\pi} \Phi_m(t)\Phi_n^*(t)\,dt = 1/2\int_{-\pi}^{\pi} \cos(0)t\,dt - 1/2\int_{-\pi}^{\pi} \cos(2m)t\,dt.$$

But $\cos(0) = 1$ and

$$\int_{-\pi}^{\pi} \cos(2mt)\,dt = 0.$$

Therefore,

$$\int_{-\pi}^{\pi} \Phi_m(t)\Phi_n^*(t)\,dt = 1/2\int_{-\pi}^{\pi} dt = 2\pi/2 = \pi.$$

If $m \neq n$ then let $m - n = p$ and $m + n = q$, where p and q are integers. Then we will have

$$\int_{-\pi}^{\pi} \Phi_p(t)\Phi_q^*(t)\,dt = 1/2\int_{-\pi}^{\pi} \cos(pt) - 1/2\int_{-\pi}^{\pi} \cos(qt)\,dt.$$

With the fact that

$$\int_{-\pi}^{\pi} \cos(2mt)\,dt = 0,$$

we have

$$\int_{-\pi}^{\pi} \Phi_p(t)\Phi_q^*(t)\,dt = 0 - 0 = 0.$$

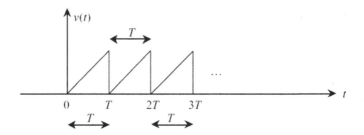

FIGURE 3.3 Periodic signal.

Finally

$$\int_{-\pi}^{\pi} \Phi_m(t) \Phi_n^*(t)\, dt = \begin{cases} \pi \text{ for } n = m \\ 0 \text{ for } n \neq m \end{cases}.$$

Therefore, the set $\Phi_m = \sin(mt)$ with $m = 1, 2, 3, \ldots$ is an orthogonal set.

3.3 PERIODIC SIGNALS

A signal $v(t)$ is periodic with a period T, if $v(t) = v(t \pm nT)$, where $n = 1, 2, 3, \ldots$. A periodic signal repeats itself every $\pm nT$ units of time. This is illustrated in Figure 3.3.

3.4 CONDITIONS FOR WRITING A SIGNAL AS A FOURIER SERIES SUM

The following conditions must be satisfied for any signal to be written as a Fourier series sum:

1. $v(t)$ must be a single-valued signal, a function of one variable.
2. There must be a finite number of discontinuities in the periodic interval.
3. There must be a finite number of maximum and minimum in the periodic interval.
4. The integral $\int_{t_0}^{t_0+T} |v(t)|\, dt$ must exist.

Any periodic function that satisfies the above four conditions can be written as a sum of basis functions.

3.5 BASIS FUNCTIONS

In the case of Fourier series representation, we will consider the set

$$\left\{ e^{\frac{j2\pi n}{T} t} \right\}_{n=-\infty}^{n=+\infty}$$

where n is an integer, as the set of basis functions.

Therefore, any signal $x(t)$ that satisfies the four conditions given earlier can be represented as a sum of these basis functions. Mathematically we write

$$x(t) = \sum_{-\infty}^{+\infty} c_n e^{\frac{j2\pi n}{T}t} \tag{3.2}$$

with $n = 0 \pm 1, \pm 2, \ldots$

Our goal is to find c_n so that $x(t)$ will be completely represented as a sum of these basis functions. Remember that the basis functions are orthogonal and

$$\int_0^T e^{\frac{j2\pi n}{T}t} e^{\frac{-j2\pi m}{T}t} \, dt = \begin{cases} T & \text{if } n = m \\ 0 & \text{if } n \neq m \end{cases}.$$

By multiplying Equation (3.2) by $e^{\frac{-j2\pi m}{T}t}$ we get

$$e^{\frac{-j2\pi m}{T}t} x(t) = \sum_{-\infty}^{+\infty} c_n e^{\frac{-j2\pi n}{T}t} e^{\frac{j2\pi n}{T}t}.$$

Now we can integrate over one period to get

$$\int_0^T e^{\frac{-j2\pi n}{T}t} x(t) \, dt = \int_0^T \sum_{-\infty}^{+\infty} c_n e^{\frac{-j2\pi m}{T}t} e^{\frac{j2\pi n}{T}t} \, dt.$$

$$\int_0^T e^{\frac{-j2\pi n}{T}t} x(t) \, dt = \sum_{-\infty}^{+\infty} c_n \int_0^T e^{\frac{-j2\pi m}{T}t} e^{\frac{j2\pi n}{T}t} \, dt.$$

Remember also that

$$\int_0^T e^{\frac{j2\pi n}{T}t} e^{\frac{-j2\pi m}{T}t} \, dt = \begin{cases} T & \text{if } n = m \\ 0 & \text{if } n \neq m \end{cases}$$

Therefore,

$$\int_0^T e^{\frac{-j2\pi n}{T}t} x(t) \, dt = T c_n$$

and

$$c_n = 1/T \int_0^T x(t) e^{\frac{-j2\pi n}{T}t} \, dt, \tag{3.3}$$

where in Equation (3.3) we set $m = n$ where $n = 0, \pm 1, \pm 2, \ldots$

3.6   THE MAGNITUDE AND THE PHASE SPECTRA

We will denote $|Cn|$ as the magnitude spectra of $x(t)$, and $\langle c_n$ as the angle spectra of $x(t)$ or the phase of $x(t)$. Notice that

$$C_{-n} = 1/T \int_0^T x(t)e^{\frac{j2\pi n}{T}t}\, dt$$

and

$$|c_{-n}| = |c_n| = \left| 1/T \int_0^T x(t)e^{\frac{j2\pi n}{T}t}\, dt \right|$$

since the magnitude of e^{jz} and the magnitude of e^{-jz} are always unity. We also see that $|c_n|$ has even symmetry, and in the same way we can deduce that $\langle c_n$ has odd symmetry.

3.7   FOURIER SERIES AND THE SIN-COS NOTATION

In terms of sin and cos functions, $x(t)$ can be expressed as

$$x(t) = a_0 + \sum_{n=1}^{\infty} a_n \cos\left(\frac{2\pi\, nt}{T}\right) + \sum_{n=1}^{\infty} b_n \sin\left(\frac{2\pi\, nt}{T}\right) \tag{3.4}$$

where

$$a_0 = c_0 = 1/T \int_0^T x(t)\, dt \tag{3.5}$$

$$a_n = 2/T \int_0^T x(t)\cos\left(\frac{2\pi\, nt}{T}\right)dt \tag{3.6}$$

$$b_n = 2/T \int_0^T x(t)\sin\left(\frac{2\pi\, nt}{T}\right)dt \tag{3.7}$$

$$c_n = 1/T \int_0^T x(t)e^{\frac{-j2\pi n}{T}t}\, dt \tag{3.8}$$

where

$$a_n = 2\mathrm{Real}(c_n)$$

and

$$b_n = -2\text{Image}(c_n),$$

where Real stands for the real part and Image stands for the imaginary part.

Example 3.2

Express the signal $v(t)$, shown in Figure 3.4, as a sum of exponential Fourier series and then in terms of the sin-cos notations.

SOLUTION

The four conditions needed to represent $v(t)$ in terms of the Fourier series are satisfied.

Therefore,

$$c_n = 1/T \int_0^T v(t) e^{\frac{-j2\pi n}{T}t} dt.$$

Substituting for $v(t)$ we get

$$c_n = 1/T \int_0^{T/2} (1) e^{\frac{-j2\pi n}{T}t} dt + 1/T \int_{T/2}^T (0) e^{\frac{-j2\pi n}{T}t} dt$$

$$c_n = 1/T \left[\frac{e^{\frac{-j2\pi n}{T}t}}{-j2\pi n/T} \right]_0^{T/2}$$

$$c_n = \left[\frac{e^{\frac{-j2\pi n}{2T}} - 1}{-j2\pi n} \right] = \left[\frac{e^{\frac{-j2\pi n}{2}}}{-j2\pi n} \right],$$

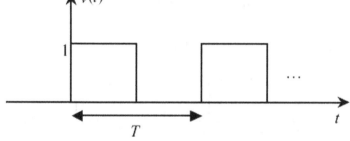

FIGURE 3.4 Signal for Example 3.2.

and finally

$$c_n = \left[\frac{e^{-n\pi} - 1}{-j2n\pi} \right].$$

Using Euler's identity we can simplify the above equation and write

$$c_n = \left[\cos(n\pi) - j\sin(n\pi) - 1 \right] / \left[-j2n\pi \right].$$

But $\sin(n\pi)$ is always zero for any integer n. Therefore c_n reduces to

$$c_n = \left[\cos(n\pi) - 1 \right] / \left[-j2n\pi \right] = \begin{cases} \dfrac{1}{jn\pi} & \text{for } n \text{ odd} \\[2mm] 0 & \text{for } n \text{ even and } n \neq 0. \\[2mm] 1/2 & \text{for } n = 0 \end{cases}$$

and the magnitude is

$$|c_n| = \begin{cases} \dfrac{1}{|n|\pi} & \text{for } n \text{ odd} \\[2mm] 0 & \text{for } n \text{ even and } n \neq 0. \\[2mm] 1/2 & \text{for } n = 0 \end{cases}$$

The phase of c_n is calculated as

$$\langle c_n = \begin{cases} \dfrac{-\pi}{2} & \text{for } n \text{ odd and negative} \\[2mm] \dfrac{\pi}{2} & \text{for } n \text{ odd and positive.} \\[2mm] 0 & \text{for } n \text{ even and zero} \end{cases}$$

The plots in Figures 3.5 and 3.6 illustrate these results.

FIGURE 3.5   Signal for Example 3.2.

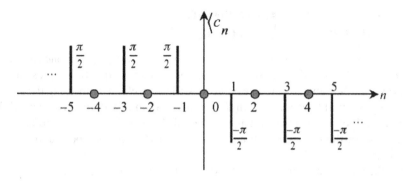

FIGURE 3.6 Signal for Example 3.2.

Now let us express $v(t)$ in terms of the sin-cos representation. We will evaluate c_n at $n = 0$ to get $a_0 = c_0 = 1/2$. We can also calculate c_0 using its defining integral as

$$c_0 = 1/T \int_0^T v(t)\, dt$$

$$= \int_0^{T/2} (1)\, dt + 1/T \int_{T/2}^T (0)\, dt = 1/2$$

$$a_n = 2Real(c_n)$$

$$= 2Real\left\{\left[\cos(n\pi) - 1\right]/\left[-j2n\pi\right]\right\}$$

$$= 2Real\left\{\left[j\cos(n\pi) - j\right]/\left[2n\pi\right]\right\} = 0$$

$$b_n = -2Imag\left\{\left[\cos(n\pi) - 1\right]/\left[-j2n\pi\right]\right\}$$

$$= -2\left[\cos(n\pi) - 1\right]/\left[2n\pi\right] = \begin{cases} \dfrac{4}{2n\pi} & \text{for } n \text{ odd} \\ 0 & \text{for } n \text{ even} \end{cases}$$

Finally, we substitute in the general form

$$x(t) = a_0 + \sum_{n=1}^{\infty} a_n \cos\left(\frac{2\pi nt}{T}\right) + \sum_{n=1}^{\infty}\left(\frac{2\pi nt}{T}\right)$$

to get the approximation to the signal $v(t)$ as

$$x(t) = 1/2 + \sum_{n \text{ odd}}\left(\frac{2}{n\pi}\sin\left(\frac{2n\pi}{T}t\right)\right).$$

3.8 FOURIER SERIES APPROXIMATION
AND THE RESULTING ERROR

In the Fourier series representation of $x(t)$, the integer n runs from one to infinity. As n approaches infinity, the series converges to the true function representation, $x(t)$. If we fix the value for the integer n, we will be, in effect, truncating the Fourier series and we will have an error in the approximation of $x(t)$, the given signal. We will take a look at the mean squared error (MSE), which uses the average difference between the truncated $x(t)$ and the original $x(t)$ over one period and square the difference.

Let the truncated $x(t)$ be given by $x_N(t)$ (we truncate at $n = N$), where

$$x_n(t) = \sum_{n=-N}^{N} d_n e^{(jnwt)}.$$

It is our objective to find d_n such that $x(t) - x_N(t)$ has the minimum mean squared value.

Let us denote the error by

$$E(t) = x(t) - x_N(t).$$

By substituting we get

$$E(t) = \sum_{n=-\infty}^{\infty} c_n e^{(jnwt)} - \sum_{n=-N}^{N} d_n e^{(jnwt)}.$$

Let

$$z_n = \begin{cases} c_n & |n| > N \\ c_n - d_n & -N < n < N \end{cases}.$$

Therefore,

$$E(t) = \sum_{n=-\infty}^{\infty} z_n e^{(jnwt)}. \tag{3.9}$$

Notice that $E(t)$ is periodic with period $T = 2\pi/w$ and the MSE is

$$\text{MSE} = \frac{1}{T} \int_{\langle T \rangle} |E(t)|^2 \, dt,$$

where we are integrating over one period. Substituting for $E(t)$ in the last integral and after some mathematical manipulations we get

$$\text{MSE} = \sum_{n=-\infty}^{\infty} |z_n|^2$$

$$\text{MSE} = \sum_{n=-N}^{N} |c_N - d_n|^2 + \sum_{|n|>N} |c_n|^2. \tag{3.10}$$

The above equation has two terms and both are positive. Therefore, to minimize the MSE we need to select $c_n = d_n$. The MSE finally is given as

$$\text{MSE} = \sum_{|n|>N} |c_N|^2. \tag{3.11}$$

3.9 THE THEOREM OF PARSEVAL

Let us take a look again at the formula we use to calculate the average power of a periodic signal

$$P_{\text{ave}} = \frac{1}{T} \int_{\langle T \rangle} |x(t)|^2 \, dt. \tag{3.12}$$

The root mean square (rms) of the average power, on the other hand, can be calculated as

$$P_{\text{ave}}(rms) = \sqrt{\frac{1}{T} \int_{\langle T \rangle} |x(t)|^2 \, dt.} \tag{3.13}$$

If $x(t) = c_n e^{jnwt}$, then

$$P_{\text{ave}} = \frac{1}{T} \int_{\langle T \rangle} |c_n e^{jnwt}|^2 \, dt$$

$$P_{\text{ave}} = \frac{1}{T} \int_{\langle T \rangle} |c_n|^2 \, dt = |c_n|^2. \tag{3.14}$$

This last equation is the Parseval equation for periodic signals.

In Example 3.2 there are two ways to calculate the average power and these two ways are presented here.

1. Using the average power formula
   The average power in this case is calculated as

$$P_{ave} = \frac{1}{T}\int_{\langle T \rangle}|v(t)|^2 = \frac{1}{T}\int_0^{T/2}|(1)|^2\, dt = 1/2.$$

2. Using the theorem of Parseval

$$P_{ave} = \sum_{n=-\infty}^{\infty}|c_n|^2$$

$$= \underset{\text{for } n=0}{\frac{(1/2)^2}{}} + \underset{\text{right side of magnitude spectra}}{\frac{(1/\pi)^2 + (1/3\pi)^2 + \cdots}{}} + \underset{\text{left side of magnitude spectra}}{\frac{(1/\pi)^2 + (/3\pi)^2 + \cdots}{}}$$

$$P_{ave} = (1/2)^2 + 2[(1/\pi)^2 + (1/3\pi)^2 + (1/5\pi)^2 + (1/7\pi)^2 + \cdots].$$

But for large enough n, P_{ave} will approach 1/2.

3.10   SYSTEMS WITH PERIODIC INPUTS

In general and in terms of time domain representation, the output of any linear time-invariant (LTI) system is given by the convolution equation

$$y(t) = \int_{-\infty}^{\infty}h(\tau)x(t-\tau)d\tau.$$

If $x(t)$ is periodic, then it can be written as

$$x(t) = e^{jwt}.$$

In this case, we will have $y(t)$ rewritten as

$$y(t) = \int_{-\infty}^{\infty}h(\tau)e^{jw(t-\tau)}d\tau = e^{jwt}\int_{-\infty}^{\infty}h(\tau)e^{-jw\tau}d\tau.$$

Let us define $H(w)$ as

$$H(w) = \int_{-\infty}^{\infty}h(\tau)e^{-jw\tau}d\tau.$$

Therefore,

$$y(t) = H(w)e^{jwt}. \tag{3.15}$$

where $H(w)$ is called the system transfer function. When we use the Fourier series representation, Equation (3.15) can be written as

$$y(t) = \sum_{-\infty}^{\infty} H(nw)c_n e^{jwt}. \tag{3.16}$$

In the above equation, we considered the input in its Fourier series representation

$$x(t) = \sum_{-\infty}^{\infty} c_n e^{jnwt.}$$

Example 3.3

Consider the following system.

$$\frac{d}{dt}y(t) + y(t) = x(t).$$

Let $x(t)$ be a periodic signal, where $x(t) = e^{jwt}$. Find $H(w)$.

SOLUTION

With $x(t) = e^{jwt}$ and $y(t) = H(w)e^{jwt}$ we have

$$\frac{d}{dt}y(t) = jwH(w)e^{jwt.}$$

Therefore, the differential equation can be written as

$$jwH(w)e^{jwt} + H(w)e^{jwt} = e^{jwt}.$$

We can drop the e^{jwt} term from both sides of the equation above (dropping the e^{jwt} term by dividing by e^{jwt} for e^{jwt} is always possible) to get

$$jwH(w) + H(w) = 1,$$

from which

$$H(w) = 1/[1 + jw],$$

which is the system transfer function of the system described by the differential equation given in this example.

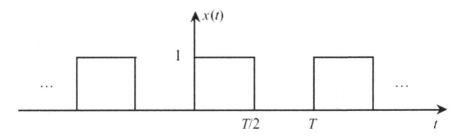

FIGURE 3.7   Signal for Example 3.4.

Example 3.4

Given $x(t)$ in Figure 3.7, with $H(nw)$ as given in Example 3.3, find $y(t)$.

SOLUTION

We found earlier that c_n for this signal is

$$c_n = \left[\cos(n\pi) - 1\right]/\left[-j2n\pi\right]$$

and

$$H(nw) = 1/\left[1 + njw\right].$$

Therefore,

$$y(t) = \sum_{-\infty}^{\infty} \left\{1/\left[1 + njw\right]\right\}\left\{\left[\cos(n\pi) - 1\right]/\left[-j2n\pi\right]\right\}e^{jnwt}.$$

3.11   A FORMULA FOR FINDING $y(t)$ WHEN $x(t)$ IS PERIODIC: THE STEADY-STATE RESPONSE

Let

$$x(t) = A\cos(wt + \phi),$$

and let us assume that we have the system transfer function $H(w)$. $y(t)$ in this case can be computed as

$$y(t) = |x(t)||H(w)|\cos(wt + \phi + \angle H(w)) \tag{3.17}$$

where | | indicates the magnitude and $\angle$ indicates the phase. Our reference in this equation is the cos angle, where ϕ is the phase angle of the input $x(t)$ and A is its amplitude.

Example 3.5

Let

$$H(w) = 1/[1 + jw],$$

and consider the input $x(t)$ to be

$$x(t) = \cos(t) - \cos(2t).$$

Find the output $y(t)$.

SOLUTION

We can use the superposition principle here and let

$$x_1(t) = \cos(t)$$

and

$$x_2(t) = \cos(2t)$$

and find $y_1(t)$ that corresponds to $x_1(t)$ and $y_2(t)$ that corresponds to $x_2(t)$. After finding $y_1(t)$ and $y_2(t)$, we can add the two outputs and get $y(t)$.

For $x_1(t)$, $w = 1$ rad/sec and we have

$$H(w = 1) = 1/[1 + j(1)]$$

and $|x_1(t)| = 1, |H(1)| = \dfrac{1}{\sqrt{2}}$, $\phi_1 = 0$, and $\angle H(1) = -45°$.

Therefore, the output $y_1(t)$ is

$$y_1(t) = |x_1(t)||H(1)|\cos(t + \phi_1 + \angle H(1))$$

$$y_1 = (1)\left(\frac{1}{\sqrt{2}}\right)\cos(t + 0 - 45°),$$

and finally

$$y_1(t) = \frac{1}{\sqrt{2}}\cos(t - 45°).$$

For $x_2(t)$, $w = 2$ rad/sec and we have

$$H(w = 2) = 1/[1 + j(2)]$$

and $|x_2(t)| = 1, |H(2)| = \dfrac{1}{\sqrt{5}}, \phi_2 = 0$, and $\angle H(2) = -63°$.

Therefore, we write

$$y_2(t) = |x_2(t)| |H(2)| \cos(2t + \phi 2 + \angle H(2))$$

$$y_2 = (1)\left(\frac{1}{\sqrt{5}}\right) \cos(2t + 0 - 63°),$$

and finally

$$y_2 = \frac{1}{\sqrt{5}} \cos(2t - 63°).$$

Now we can add the two solutions $y_1(t)$ and $y_2(t)$ (notice the minus sign before $x_2(t)$) to get

$$y(t) = \frac{1}{\sqrt{2}} \cos(t - 45°) - \frac{1}{\sqrt{5}} \cos(2t - 63°).$$

3.12   SOME INSIGHT: WHY THE FOURIER SERIES

We will discuss some of the many uses of the Fourier series.

3.12.1   NO EXACT SINUSOIDAL REPRESENTATION FOR $x(t)$

As we have seen in this chapter, the Fourier series representation can be obtained for periodic signals with finite discontinuities and bounded magnitudes. In cases where the inputs to the systems that we study in real life have such shapes, periodicity, and magnitudes, we will use Equation (3.17) to find the system output $y(t)$. We will prefer to write these input signals as a sum of sinusoids.

If we can write $x(t)$ as a sum of sin and cos terms in a compact mathematical form, yet closely approximate the actual $x(t)$ form, we can find the corresponding $y(t)$, which should closely match the true output. This is one of the advantages the Fourier series offers.

3.12.2   THE FREQUENCY COMPONENTS

The other advantage is the information we get from c_n, the Fourier series coefficient. We are able to identify the frequency contents of the input signal $x(t)$, both the magnitude and the phase spectrum. Knowing the frequency contents of the input signal gives us more insights into the way we analyze and design linear systems; we will know what frequencies will pass and what frequencies will not pass through a particular system.

3.13   END-OF-CHAPTER EXAMPLES

EOCE 3.1

Evaluate each of the following expressions:

1. $\dfrac{1+j}{1-j}+\dfrac{1+2j}{1-2j}-1+j$

2. $\dfrac{1+j}{1-j}-\dfrac{3+j}{1-j}+\dfrac{1+j}{2-j}$

3. $\dfrac{1+j}{1-j}+\dfrac{1+2j}{1-2j}-1+j$

Use MATLAB in the calculations.

SOLUTION

Every complex number has a magnitude and an angle. At the MATLAB prompt, we type the following commands for expression 1.

```
Expression_1 = (1 + j)/(1 - j) + (1 + 2*j)/(1 - 2*j)
```

to get

```
Expression_1 =
-0.6000 + 1.8000i
Magnitude_of_Expression_1 = abs(Expression_1)
```

to get

```
Magnitude_of_Expression_1 =
1.8974
Angle_of_Expression_1 = angle(Expression_1)
```

to get

```
Angle_of_Expression_1 =
1.8925
```

For expression 2 we type the commands

```
Expression_2 = (1 + j)/(1 - j) - (3 + j)/(1 - j) + (1 + j)/(2 - j)
```

to get

```
Expression_2 =
-0.8000 - 0.4000i
Magnitude_of_Expression_2 = abs (Expression_2)
```

to get

```
Magnitude_of_Expression_2 =
0.8944
```

```
Angle_of_Expression_2 = angle(Expression_2)
```

to get

```
Angle_of_Expression_2 =
-2.6779
```

For expression 3 we type the commands at the prompt

```
Expression_3 = (1 + j)/(1 - j) + (1 + 2*j)/(1 - 2*j) - 1 + j
```

to get

```
Expression_3 =
-1.6000 + 2.8000i
Magnitude_of_Expression_3 = abs (Expression_3)
```

to get

```
Magnitude_of_Expression_3 =
3.2249
Angle_of_Expression_3 = angle(Expression_3)
```

to get

```
Angle_of_Expression_3 =
2.0899
```

EOCE 3.2

In Example 3.2 of this chapter we worked following problem: Express the signal $v(t)$, shown again in Figure 3.8, as a sum of exponential Fourier series and then use the sin-cos notations to express the same signal. The final calculation for c_n was

$$c_n = \left[\cos(n\pi) - 1\right]/\left[-j2n\pi\right],$$

which is a complex number that has a magnitude and an angle.

Use MATLAB to show that the plots obtained before for the magnitude and the angle for c_n are accurate.

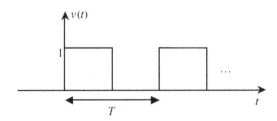

FIGURE 3.8 Signal for EOCE 3.2.

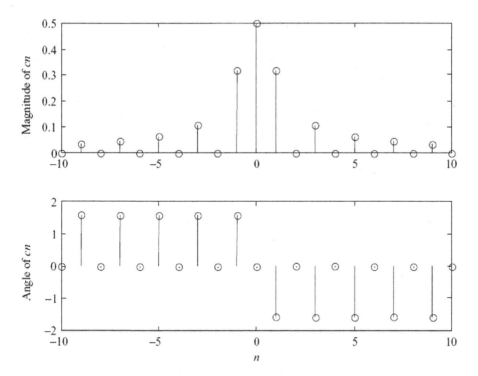

FIGURE 3.9 Plots for EOCE 3.2.

SOLUTION

We use the MATLAB script EOCE3_21 and the plots are shown in Figure 3.9.

For $T = 1$, we can plot an approximation to the given signal. In Example 3.2, we have calculated $x(t)$ as

$$x(t) = 1/2 + \sum_{n \text{ odd}} \left(\frac{2}{n\pi} \sin\left(\frac{2n\pi}{T} t \right) \right).$$

Using MATLAB, we can plot $x(t)$ to see how it can approximate $v(t)$ by writing the script EOCE3_22.

The plots are shown in Figure 3.10.

EOCE 3.3

Consider the signal $x(t)$ shown in Figure 3.11 as an input to the linear first-order system described by the differential equation

$$\frac{d}{dy} y + y = x(t)$$

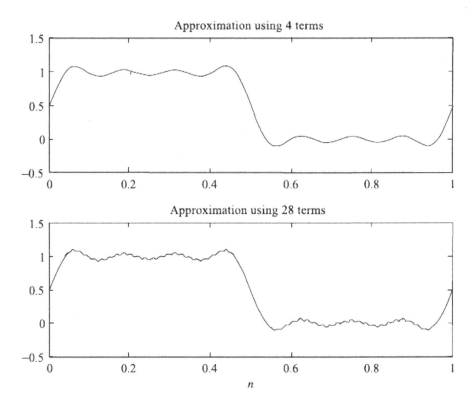

FIGURE 3.10 Plots for EOCE 3.2.

1. Find an approximation to $x(t)$.
2. Plot the magnitude and phase of c_n.
3. Plot the approximation to $x(t)$.
4. Find the output to the system with $n = 2$ and plot it.
5. Find the average power in the time and frequency domains.

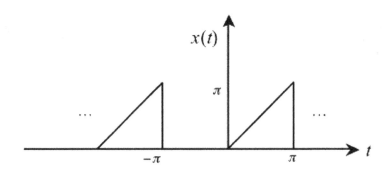

FIGURE 3.11 Signal for EOCE 3.3.

SOLUTION

1. Mathematically, $x(t)$ can be written as

$$x(t) = \begin{cases} 0 & -\pi < t < 0 \\ t & 0 < t < \pi \end{cases}$$

with period 2π.

The Fourier coefficients are determined as in the following.

$$a_0 = \frac{1}{2\pi} \int_{-\pi}^{\pi} x(t)\, dt.$$

By substituting for $x(t)$ we arrive at

$$a_0 = \frac{1}{2\pi} \int_{-\pi}^{\pi} t\, dt = \frac{\pi}{4}$$

and

$$a_n = \frac{1}{2\pi} \int_{-\pi}^{\pi} x(t)\cos(nt)\, dt.$$

Substituting again for $x(t)$ we obtain

$$a_n = \frac{1}{\pi} \int_{-\pi}^{0} 0\cos(nt)\, dt + \frac{1}{\pi} \int_{0}^{\pi} t\,\cos(nt)\, dt.$$

Using the method of integrating by parts, we will get

$$a_n = \frac{1}{\pi}\left[\frac{t}{n}\sin(nt) + \frac{1}{n^2}\cos(nt) \right]_0^{\pi} = \frac{1}{n^2\pi}\left[\cos(n\pi) - 1 \right]$$

for $n = 1, 2, 3, \ldots$

We can find the other coefficient as in the following:

$$b_n = \frac{1}{\pi} \int_{-\pi}^{\pi} x(t)\sin(nt)\, dt.$$

With $x(t)$ substituted we get

$$b_n = \frac{1}{\pi} \int_{-\pi}^{\pi} 0\sin(nt)\, dt + \frac{1}{\pi} \int_{0}^{\pi} t\,\sin(nt)\, dt.$$

By carrying out the integration we get

$$b_n = \frac{1}{c}\left[2\frac{t}{n}\cos(nt) + \frac{1}{n^2}\sin(nt)\right]_0^\pi = -\frac{1}{2}\left[\cos(n\pi)\right]$$

for $n = 1, 2, 3, \ldots$

By observing that $(-1)^n = \cos(n\pi)$ and that $a_n = 0$ for even n, we can write the approximation for $x(t)$ as

$$x(t) = \frac{\pi}{4} - \frac{2}{\pi}\sum_1^\infty \frac{\cos(2n-1)t}{(2n-1)^2} - \sum_1^\infty (-1)^n \frac{\sin(nt)}{n}.$$

2. We can plot the magnitude and the phase of

$$c_n = \frac{1}{2n^2\pi}\left[(-1)^n - 1\right] + j\frac{1}{2n}(-1)^n$$

for $n = 0, \pm1, \pm2, \ldots$ by writing the Matlab script EOCE3_31.

The plots are shown in Figure 3.12.

3. Let us now use MATLAB to plot the approximation, $x(t)$, to the triangular wave given using the script EOCE3_32.

The plots are shown in Figure 3.13.

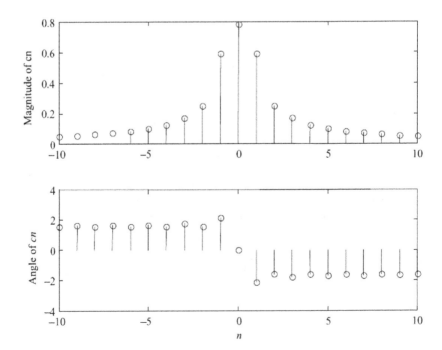

FIGURE 3.12 Plots for EOCE 3.3.

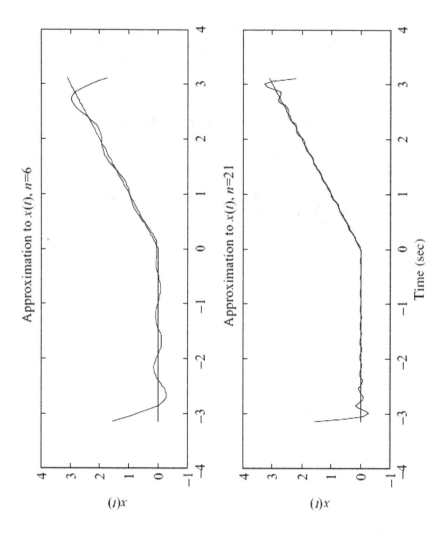

FIGURE 3.13   Plots for EOCE 3.3.

4. If we take $n = 2$, the input $x(t)$ will be

$$x(t) = \frac{\pi}{4} - \frac{2\cos(t)}{\pi} - \frac{2\cos(3t)}{9\pi} + \sin(t) - \frac{\sin(2t)}{2}.$$

Writing terms as cos terms only so that our reference will be the cos function, $x(t)$ becomes

$$x(t) = \frac{\pi}{4} - \frac{2\cos(t)}{\pi} - \frac{2\cos(3t)}{9\pi} + \cos\left(t - \frac{\pi}{2}\right) - \frac{\cos\left(2t - \frac{\pi}{2}\right)}{2}.$$

This input can be divided into five terms, $x_1(t)$, to $x_5(t)$.
The transfer function of the system $H(jw)$ is determined as

$$H(jw) = \frac{1}{jw + 1}.$$

We will use MATLAB commands plus the formula for calculating $y(t)$ if the input is sinusoidal in the following.
For the first term, $x_1(t) = \pi/4$, $w = 0$, and $H_1(jw) = 1$.

```
magnitude_of_x1 = abs(pi/4)
```

to get

```
magnitude_of_x1 =
0.7854
phase_of_x1 =angle(pi/4)
```

to get

```
phase_of_x1 =
0
magnitude_of_H1 = abs(1)
```

to get

```
magnitude_of_H1 =1
phase_of_H1 =angle(1)
```

to get

```
phase_of_H1 =
0
```

Therefore,

$$y_1(t) = .7854$$

For the second term, $x_2(t) = 2(\cos(t))/\pi$, $w = 1$, and $H_2(jw) = 1/(1 + j)$.

The magnitude of $x_2(t) = 2/\pi$ and the phase is 0.

```
magnitude_of_H2 = abs(1/(1 + j))
```

to get

```
magnitude_of_H2 =
0.7071
phase_of_H2 = angle(1/(1 + j))
```

to get

```
phase_of_H2 =
-0.7854
```

Therefore,

$$y_1(t) = (2/\pi)(.7071)\cos(t - .7854)$$

For the third term, $x_3(t) = -2/(9\pi)\cos(3t)$, $w = 3$, and $H_3(jw) = 1/(1 + 3j)$. The magnitude of $x_3(t) = 2/(9\pi)$ and the phase is $= 0$.

```
magnitude_of_H3 = abs(1/(1 = 3*j))
```

to get

```
magnitude_of_H3 =
3162
phase_of_H3 = angle(1/(1 + 3*j))
```

to get

```
phase_of_H3 =
-1.2490
```

Therefore,

$$y_3(t) = [2/(9\pi)][.3162]\cos(3t - 1.249).$$

For the fourth term, $x_4(t) = \cos(t - \pi/2)$. The phase of $x_4(t) = -\pi/2$, $w = 1$, $|H_4(jw)| = |H_2(jw)| = .7071$, and phase of $H_4(jw)$ is the same as the phase of $H_2(jw) = -.7854$ rad.
Therefore,

$$y_4(t) = .7071\cos(t - \pi/2).$$

For the fifth term, $x_5(t) = \cos(2t - \pi/2)$, $w = 2$, and $H_4(jw) = 1/(1 + 2j)$. The magnitude of $x_5(t) = 1$ and the phase is $-\pi/2$.

```
magnitude_of_x5 = abs(1/(1 + 2*j))
```

to get

```
magnitude_of_x5 =
0.4472
phase_of_H5 = angle(1/(1 + 2*j))
```

to get

```
phase_of_H5 =
-1.1071
```

Therefore,

$$y_5(t) = (.4472)(1)\cos(2t - \pi/2 - 1.1071).$$

The total solution is

$$y(t) = y_1(t) + y_2(t) + y_3(t) + y_4(t) + y_5(t).$$

By substituting for the individual outputs we arrive at

$$(t) = .7854 - (2/\pi)(.7071)\cos(t - .7854) - \left[2/(9\pi)\right][.3162]$$

$$\times \cos(3t - 1.249) + .7071\cos(t - \pi/2)$$

$$-(.4472)(1)\cos(2t - \pi/2 - 1.1071).$$

Let us now plot the response for $n = 1$ and $n = 2$. We use MATLAB to do that by writing the script EOCE3_33.
The plot is shown in Figure 3.14.
5. The average power in the signal is calculated as in the following. In the time domain, the average power is

$$P_{ave} = \frac{1}{T}\int_T |x(t)|^2\, dt.$$

By substituting for $x(t)$ we get

$$P_{ave} = \frac{1}{2\pi}\int_{-\pi}^{\pi} |x(t)|^2\, dt = \frac{1}{2\pi}\int_{-\pi}^{0} 0\, dt + \frac{1}{2\pi}\int_{0}^{\pi} t^2\, dt.$$

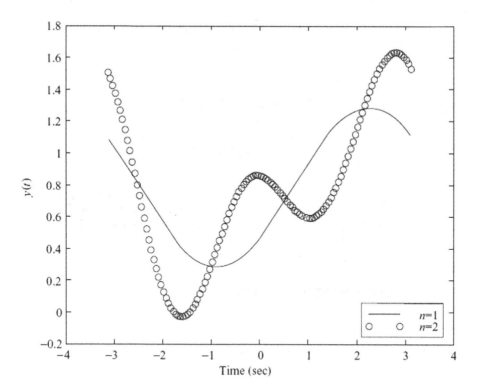

FIGURE 3.14   Plots for EOCE 3.3.

By carrying out the integration we get

$$P_{ave} = \frac{1}{2\pi} \int_0^\pi t^2 dt \frac{1}{2\pi} \left[\frac{t^3}{3} \right]_0^\pi = \frac{\pi^2}{6}.$$

In the frequency domain, the average power is

$$P_{ave} = \sum_{-\infty}^{\infty} |c_n|^2$$

and

$$c_n = \frac{1}{2\pi^2 \pi} \left[\cos(n\pi) - 1 \right] + j \left[\frac{1}{2n} \cos(n\pi) \right].$$

Remember that $|c_n|$ is even symmetric and $c_0 = \pi/4$. The average power becomes

$$P_{ave} = \frac{\dfrac{\pi^2}{16}}{\text{term for } n = 0} + \frac{2}{2-\text{sided spectra}}$$

$$\times \left[\frac{1}{\pi^2} + \frac{1}{4} + \frac{1}{16} + \frac{1}{3^4 \pi^2} + \frac{1}{4(9)} + \frac{1}{64} + \cdots \right].$$

Using the above equation, $p_{ave} = 1.5338$ and is obtained by only considering five terms. The actual average power is 1.6449. We know that the Parseval theorem, for large n, will result in an average power that is very close to 1.6449. The error in this example is about 11%.

3.14 END-OF-CHAPTER PROBLEMS

EOCP 3.1

Perform the following complex arithmetic:

1. $\dfrac{1+2j}{1-j} + 1$

2. $\dfrac{-2j}{1-j} - \dfrac{1}{2j} + j$

3. $\dfrac{1+2j}{j} + 1 - j$

4. $1 + 10j - \dfrac{1+10j}{1-j}$

5. $\dfrac{1+10j}{1-10j} + \dfrac{1+10j}{1-j} - 10$

EOCP 3.2

Put the following complex numbers in polar form

1. $\dfrac{1-j}{1+j} + 2$

2. $\dfrac{-j}{1+j} + j + 1$

3. $\dfrac{1-2j}{-j} - 1 + j$

4. $-j + \dfrac{-j}{1-j}$

5. $\dfrac{-1+3j}{-1+10j} + \dfrac{1-j}{-1-j}$

EOCP 3.3

The angles are given in degrees. Put the following complex numbers in rectangular form.

1. $10\langle 90$

2. $10\langle 90 + \dfrac{1\langle 120}{1\langle -30}$

3. $10\langle -90 + \dfrac{(2\langle 0)(4\langle 120)}{1\langle -30} - 1\langle -120$

EOCP 3.4

Consider the signal shown in Figure 3.15.

1. Find approximation to $x(t)$.
2. Find and plot the magnitude and phase spectra.
3. Find the average power using $x(t)$ directly.
4. Use the Fourier coefficients to find the average power.

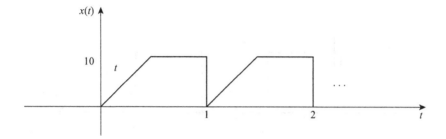

FIGURE 3.15   Signal for EOCP 3.4.

EOCP 3.5

Find the system transfer function for each of the following systems.

1. $\dfrac{d}{dt}y(t)+5y(t)=x(t)$

2. $3\dfrac{d}{dt}y(t)+3y(t)=x(t)+3\dfrac{d}{dt}x(t)$

3. $\dfrac{d^2}{dt}y(t)-1/2y(t)=x(t)$

4. $\dfrac{d^2}{dt}y(t)+1/2\dfrac{d}{dt}y(t)=x(t)+\dfrac{d}{dt}x(t)$

5. $\dfrac{d^2}{dt}y(t)+1/2\dfrac{d}{dt}y(t)+y(t)=x(t)$

EOCP 3.6

Consider the systems with the input $x(t)$ as represented in Figure 3.16.

1. $\dfrac{d}{dt}y(t)+5y(t)=x(t)$

2. $3\dfrac{d}{dt}y(t)+3y(t)=x(t)+3\dfrac{d}{dt}x(t)$

Take only three terms from the approximated $x(t)$ to find the output for the first system. Take only two terms from the approximated $x(t)$ to find the output for the last system.

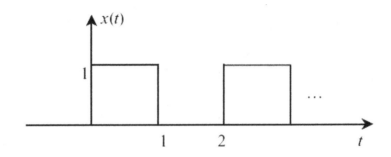

FIGURE 3.16   Signal for EOCP 3.6.

REFERENCES

Cogdell, J.R. *Foundations of Electrical Engineering*, 2nd ed., Englewood Cliffs, NJ: Prentice-Hall, 1996.

Denbigh, P. *System Analysis and Signal Processing*, Reading, MA: Addison-Wesley, 1998.

Golubitsky, M. and Dellnitz, M. *Linear Algebra and Differential Equations Using MATLAB*, Stamford, CT: Brooks/Cole, 1999.

Harman, T.L., Dabney, J., and Richert, N. *Advanced Engineering Mathematics with MATLAB*, Stamford, CT: Brooks/Cole, 2000.

Moore, H. *MATLAB for Engineers*, Upper Saddle River, NJ: Prentice Hall, 2007.

Nilson, W.J. and Riedel, S.A. *Electrical Circuits*, 6th ed., Englewood Cliffs, NJ: Prentice-Hall, 2000.

Phillips, C.L. and Parr, J.M. *Signals, Systems, and Transforms*, 2nd ed., Englewood Cliffs, NJ: Prentice-Hall, 1999.

Strum, R.D. and Kirk, D.E. *Contemporary Linear Systems*, Boston, MA: PWS, 1996.

The MathWorks. *The Student Edition of MATLAB*, Englewood Cliffs, NJ: Prentice-Hall, 1997.

Wylie, R.C. and Barrett, C.L. *Advanced Engineering Mathematics*, 6th ed., New York, NY: McGraw-Hill, 1995.

Ziemer, R.E., Tranter, W.H., and Fannin, D.R. *Signals Systems Continuous and Discrete*, 4th ed., Englewood Cliffs, NJ: Prentice-Hall, 1998.

4 The Fourier Transform and Linear Systems

4.1 DEFINITION

The Fourier transform is a frequency-domain transform that makes solution, design, and analysis of linear systems simpler. It also gives some insights about the frequency contents of signals where these insights are harder to see in real-time systems. There are other important uses for the Fourier transform, but we will concentrate only on the issues introduced in this definition.

4.2 INTRODUCTION

We needed the Fourier series expansion to approximate signals that are periodic in time. In this chapter, we will utilize Fourier series to try to express nonperiodic signals using what we call the Fourier transform. The Fourier transform is the frequency-domain representation of nonperiodic time-domain signals. Consider the time-domain periodic signal $x_p(t)$ in Figure 4.1.

Let us fix the size of the pulse (the height and the width) as seen in the graph in Figure 4.2. The height of the pulse is a, and the width of the pulse is τ units of time. Imagine now that the period T is extended to infinity. In this case, our signal in Figure 4.1 will look like the signal in Figure 4.2.

You can see that this signal looks like a nonperiodic signal. We also know that $w = 2\pi/T$. As $T \to \infty$, $w \to 0$. As w becomes very small, $\frac{w \to 0}{2\pi}$ can be written as $\frac{dw}{2\pi}$, where dw is a very small differential. In terms of Fourier series we write

$$x_p(t) = \sum_{n=-\infty}^{\infty} c_n e^{jnwt} \tag{4.1}$$

and the coefficients are

$$c_n = \frac{1}{T} \int_{\frac{-T}{2}}^{\frac{T}{2}} x_p(t) e^{-jnwt} dt. \tag{4.2}$$

As $T \to \infty$, nw will be a continuous variable (w will be small). Substitute $dw/2\pi$ for $1/T$ in the equation for c_n above to get

$$\frac{c_n}{dw} = \frac{1}{2\pi} \int_{-\infty}^{+\infty} x_p(t) e^{-jwt} dt. \tag{4.3}$$

FIGURE 4.1 Periodic signal.

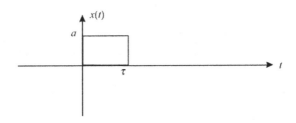

FIGURE 4.2 Signal when period is infinite.

In the equation for $x_p(t)$ above and in the limit as seen in the graph for $x(t)$, $x_p(t)$ becomes $x(t)$ and the summation becomes an integration. In this case

$$x(t) = \int_{-\infty}^{+\infty} \left[\underbrace{\left[\int_{-\infty}^{+\infty} x(t)e^{-jwt}\, dt \right]}_{\text{this whole term is function of } w,\ \text{call it } X(w)} \right] e^{jwt}\, \frac{dw}{2\pi}. \qquad (4.4)$$

Therefore,

$$x(t) = \frac{1}{2\pi} \int_{-\infty}^{\infty} X(w)e^{jwt}\, dw. \qquad (4.5)$$

4.3 THE FOURIER TRANSFORM PAIRS

In general, for any input $x(t)$ that satisfies the conditions

1. $\displaystyle\int_{-\infty}^{\infty} |x(t)|\, dt < \infty$

2. $x(t)$ has bounded variations

$x(t)$ can be transformed into a domain called the Fourier transform domain. The transform pairs that take $x(t)$ from the time domain to $X(w)$ in the Fourier frequency domain, and vice versa, are

$$X(w) = \int_{-\infty}^{\infty} x(t)e^{-jwt}\,dt \tag{4.6}$$

$$x(t) = \frac{1}{2\pi}\int_{-\infty}^{\infty} X(w)e^{jwt}\,dw. \tag{4.7}$$

These are referred to as the Fourier transform pairs (Table 4.1). If $X(w)$ is the Fourier transform of $x(t)$ and $x(t)$ is the inverse transform of $X(w)$, then we can use the notation

$$x(t) \Leftrightarrow X(w)$$

to indicate the two-way transform from time to frequency and from frequency back to time.

TABLE 4.1
Fourier Transform Table: Selected Pairs

	$x(t)$	$X(w)$
1.	$u(t)$	$\pi\delta(w) + \dfrac{1}{jw}$
2.	$\delta(t)$	1
3.	1	$2\pi\delta(w)$
4.	$e^{jw_0 t}$	$2\pi\delta(w - w_0)$
5.	$\cos(w_0 t)$	$\pi[\delta(w - w_0) + \delta(w + w_0)]$
6.	$\sin(w_0 t)$	$\pi/j[\delta(w - w_0) - \delta(w + w_0)]$
7.	$\cos(w_0 t)u(t)$	$\pi/2\big[\delta(w - w_0) + \delta(w + w_0)\big] + \dfrac{jw}{w_0^2 - w^2}$
8.	$\sin(w_0 t)u(t)$	$\pi/2j\big[\delta(w - w_0) + \delta(w + w_0)\big] + \dfrac{w_0}{w_0^2 - w^2}$
9.	$e^{-at}u(t)\ a > 0$	$\dfrac{1}{a + jw}$
10.	$te^{-at}u(t)\ a > 0$	$\left(\dfrac{1}{a + jw}\right)^2$
11.	$\dfrac{t^{n-1}}{(n-1)!}e^{-at}u(t)\,a > 0$	$\left(\dfrac{1}{a + jw}\right)^n$
12.	$e^{-a|t|}u(t)\ a > 0$	$\dfrac{2a}{a^2 + w^2}$
13.	$e^{-at^2}u(t)\,a > 0$	$\sqrt{\dfrac{\pi}{a}}\,e^{\frac{-w^2}{4a}}$
14.	$\displaystyle\sum_{n=-\infty}^{\infty}\delta(t - nT)$	$\dfrac{2\pi}{T}\displaystyle\sum_{n=-\infty}^{\infty}\delta\left(w - \dfrac{2n\pi}{T}\right)$

Example 4.1

Find the Fourier transform of $x(t) = \delta(t)$.

SOLUTION

Using the defining integral, the Fourier transform equation that takes $x(t)$ from its time domain and puts it in its frequency domain is

$$X(w) = \int_{-\infty}^{\infty} x(t)e^{-jwt}dt = \int_{-\infty}^{\infty} \delta(t)e^{-jwt}dt.$$

Using the sifting property of the impulse signal we can simplify to get

$$X(w) = \left| e^{-jw(0)} \right| = 1.$$

Therefore, we can write using the established notation

$$x(t) = \delta(t) \Leftrightarrow X(w) = 1.$$

Example 4.2

Find the Fourier transform of $sgn(t)$.

SOLUTION

Let us take the derivative of $sgn(t)$, the signal that was presented in Chapter 1, and write

$$\frac{d}{dt} sgn(t) = 2\delta(t).$$

Let us take the Fourier transform on both sides and use the Fourier transform derivative property as shown in Table 4.2 with $x(t) = sgn(t)$ to get

$$jwX(w) \Leftrightarrow 2(1).$$

This last expression says that the Fourier transform of $sgn(t)$ is $2/jw$ or

$$sgn(t) \Leftrightarrow 2/jw.$$

Example 4.3

Find the Fourier transform of $x(t) = e^{jw_0 t}$ and then deduce the Fourier transform of the constant signal $x(t) = k$.

SOLUTION

Consider an impulse signal in the frequency domain as

$$X(w) = \delta(w - w_0).$$

TABLE 4.2
Properties of the Fourier Transform

	Time Signal	Fourier Transform Signal
1.	$x(t)$	$X(w)$
2.	$\displaystyle\sum_{n=1}^{N} a_n x_n(t)$	$\displaystyle\sum_{n=1}^{N} a_n X_n(w)$
3.	$x(-t)$	$X(-w)$
4.	$x(t - t_0)$	$X(w)e^{-jwt_0}$
5.	$x(\alpha t)$	$\dfrac{1}{|\alpha|} X(w/\alpha)$
6.	$\dfrac{d^n}{dt} x(t)$	$(jw)^n X(w)$
7.	$x(t) * h(t)$	$X(w)H(w)$
8.	$x(t)h(t)$	$\dfrac{1}{2\pi} X(w) * H(w)$
9.	$x(t)e^{jw_0 t}$	$X(w - w_0)$
10.	$\displaystyle\int_{-\infty}^{\infty} x(\tau)d\tau$	$\dfrac{X(w)}{jw} + \pi X(0)\delta(w)$
11.	$x\left(\dfrac{t}{a}\right)$	$\dfrac{1}{|a|} X(w/a)$

The inverse transform of $X(w)$ is

$$x(t) = \frac{1}{2\pi} \int_{-\infty}^{\infty} \delta(w - w_0)e^{jwt}dt.$$

By using the sifting property of the impulse function as explained in Chapter 1, we can write

$$x(t) = \frac{1}{2\pi} e^{jw_0 t}.$$

But $x(t)$ is the inverse transform of $\delta(w - w_0)$, therefore

$$e^{jw_0 t} \Leftrightarrow 2\pi\delta(w - w_0).$$

Now let $w_0 = 0$. The above transform becomes

$$1 \Leftrightarrow 2\pi\delta(w).$$

Now we can use the linearity property from Table 4.2 and deduce

$$k \Leftrightarrow 2\pi k\delta(w).$$

Example 4.4

Consider the following signal $x(t)$ of period T, where $w = 2\pi/T$. We can write $x(t)$ using the Fourier series as

$$x(t) = \sum_{-\infty}^{\infty} c_n e^{jnw_0 t}.$$

Let $\Im$ represent the Fourier transform operator. What is the $\Im(x(t))$?

SOLUTION

The Fourier transform of $x(t)$ is

$$X(w) = \sum_{-\infty}^{\infty} c_n \Im\left[e^{jnw_0 t}\right],$$

which simplifies to

$$X(w) = \sum_{-\infty}^{\infty} c_n\left[2\pi\delta(w - nw_0)\right].$$

$X(w)$ can be viewed as a train of impulses on the w axis having values only at nw_0, where $n = 0, \pm1, \pm2, \pm3, \ldots$. These impulses are separated from each other by w_0.

We can also find the magnitude of $X(w)$ as

$$\left|X(w)\right| = \left|\sum_{-\infty}^{\infty} c_n \Im\left[e^{jnw_0 t}\right]\right|$$

or

$$\left|X(w)\right| = 2\pi|c_n|.$$

Example 4.5

Consider the following signal with period T.

$$x(t) = \sum_{-\infty}^{\infty} \delta(t - nT)$$

Find the Fourier transform of the given signal.

SOLUTION

Let us first write the given signal as a sum of Fourier series. We can write the coefficients first as

$$c_n = \frac{1}{T} \int_{\langle T \rangle} x(t) e^{-j\frac{2\pi nt}{T}} dt.$$

With $x(t)$ as the given train of impulses, we can utilize the sifting property for integrals involving impulses to get

$$c_n = \frac{1}{T} \int_{\langle T \rangle} x(t) e^{-j\frac{2\pi nt}{T}} dt = \frac{1}{T}.$$

With $x(t) = \sum_{-\infty}^{\infty} \delta(t - nT)$, the signal can be written as

$$x(t) = \sum_{-\infty}^{\infty} \frac{1}{T} \left[e^{\frac{j2\pi nt}{T}} \right],$$

from which

$$X(w) = \sum_{-\infty}^{\infty} \frac{1}{T} \Im \left[e^{\frac{j2\pi nt}{T}} \right],$$

or finally

$$X(w) = \frac{2\pi}{T} \sum_{n=-\infty}^{\infty} \delta\left(w - \frac{2n\pi}{T} \right).$$

In this last equation, we used entry number 4 in Table 4.1 for the Fourier transform pairs.

Example 4.6

Let $x_1(t) = \delta(t)$ and $x_2(t) = e^{jw_0 t}$. Find the Fourier transform of $(3x_1(t) + 5x_2(t))$.

SOLUTION

We can use the second entry in Table 4.2 to transform the indicated sum. $x_1(t)$ can be transformed using Table 4.1 as

$$x_1(1) \Leftrightarrow 1.$$

With the help of Table 4.1 we can also write

$$x_2(t) \Leftrightarrow 2\pi\delta(w - w_0),$$

and therefore,

$$3x_1(t) + 5x_2(t) \Leftrightarrow 3(1) + 5(2)\pi\delta(w - w_0) = 3 + 10\pi\delta(w - w_0).$$

Example 4.7

Find the Fourier transform of $x(t) = 1 + 2u(t)$.

SOLUTION

From entries 3 and 1 in Table 4.1 and entry 2 in Table 4.2 (the linearity property) we see that

$$1 \Leftrightarrow 2\pi\delta(w)$$

and

$$2u(t) \Leftrightarrow 2\left[\pi\delta(w) + 1/jw\right].$$

Therefore,

$$x(t) \Leftrightarrow X(w) = 4\pi\delta(w) + 2/jw.$$

Example 4.8

Find the Fourier transform of $x(t - 2)$, where $x(t)$ is as given in Example 4.7 above.

SOLUTION

Since in Example 4.7 we had $x(t) = 1 + 2u(t)$, therefore,

$$x(t - 2) = 1 + 2u(t - 2).$$

Using the time-shifting property, entry 4 in Table 4.2, we write

$$x(t - 2) \Leftrightarrow X(w)e^{-2jw}.$$

Therefore,

$$x(t - 2) = 1 + 2u(t - 2) \Leftrightarrow X(w)e^{-2jw} = \left[4\pi\delta(w) + 2/jw\right]e^{-2jw}.$$

Example 4.9

Find the Fourier transform of $\dfrac{d}{dt}x(t)$, where $x(t)$ is as given in Example 4.7.

SOLUTION

FIRST METHOD

From entry 6, Table 4.2, with $n = 1$, we have

$$\frac{d}{dt}x(t) \Leftrightarrow jwX(w) = jw\left[4\pi\delta(w) + 2/jw\right].$$

Remember that with any function $f(t)$, $f(t)\delta(t) = \delta(t)\,f(t = 0)$. Therefore,

$$\frac{d}{dt}x(t) \Leftrightarrow jwX(w) = jw\left[4\pi\delta(w) + 2/jw\right] = 2.$$

SECOND METHOD

Let us differentiate $x(t)$ first and write

$$\frac{d}{dt}x(t) = \frac{d}{dt}\left[1 + 2u(t)\right] = 2\delta(t),$$

where the derivative of $u(t)$ is $\delta(t)$.
   Therefore,

$$2\delta(t) \Leftrightarrow 2(1) = 2.$$

Example 4.10

Find the Fourier transform of $x(t/3)$, where $x(t)$ is as in Example 4.7.

SOLUTION

Using entry 5, Table 4.2, we have

$$x(t/3) \Leftrightarrow \left[1/|1/3|\right]X(w/1/3).$$

By substituting for $x(t/3)$ and $X(w/(1/3))$ we arrive at

$$\left[1 + 2u(t/3)\right] \Leftrightarrow \left[1/|1/3|\right]\left[4\pi\delta(w/(1/3)) + 2/j(w/(1/3))\right]$$

$$= 3\left[4\pi\delta(3w) + 6/jw\right].$$

Example 4.11

Find the Fourier transform of x(t)*x(t – 2), where * means convolution and x(t) is as given in Example 4.7.

SOLUTION

From entry 7, Table 4.2, we have

$$x(t)*x(t-2) \Leftrightarrow X(w)X(w)e^{-2jw}.$$

If we substitute for x(t) and x(t – 2), we have

$$[1+2u(t)]*[1+2u(t-2)] \Leftrightarrow [4\pi\delta(w)+2/jw][4\pi\delta(w)+2/jw]e^{-2jw}.$$

Example 4.12

Let x(t) = e⁻ᵗu(t), and the system's impulse response, h(t) = δ(t).

1. Find the Fourier transform of x(t) and h(t).
2. Let y(t) be the output of a linear time-invariant system, with x(t) and h(t), as given in this example. Find the output y(t).
3. Let x(t) = u(t) and h(t) = e⁻ᵗu(t). Find y(t), the output, for this new system.

SOLUTION

1. From Table 4.1, entry 9, we have

$$x(t) = e^{-t}u(t) \Leftrightarrow X(w) = 1/[1+jw],$$

and from Table 4.1, entry 2, we also have

$$h(t) = \delta(t) \Leftrightarrow H(w) = 1.$$

2. From Table 4.2, entry 7, we have

$$y(t) = x(t)*h(t) \Leftrightarrow Y(w) = X(w)H(w).$$

Therefore,

$$Y(w) = (1)[1/(1+jw)] = [1/(1+jw)].$$

We can also look at Table 4.1 again and see the inverse transform of Y(w), which is

$$y(t) = e^{-t}u(t).$$

3. With $x(t) = u(t)$ and $h(t) = e^{-t}u(t)$, the output $y(t)$ is

$$y(t) = x(t) * h(t) = u(t) * e^{-t}u(t).$$

We can take the Fourier transform of this last equation to get

$$Y(w) = \left[\frac{1}{jw} + \pi\delta(w)\right]\left[\frac{1}{1+jw}\right]$$

$$Y(w) = \left[\frac{\pi\delta(w)}{1+jw}\right] + \left[\frac{1}{1+jw}\right]\left[\frac{1}{jw}\right].$$

Always remember that with any function $f(t)$, $f(t)\delta(t) = \delta(t)f(t=0)$. With this note and using partial fractions we write

$$Y(w) = \left[\frac{\pi\delta(w)}{1}\right] + \left[\frac{1}{w(j-w)}\right].$$

We can do partial fraction expansion on $\dfrac{1}{w(j-w)}$ and write

$$Y(w) = \left[\frac{\pi\delta(w)}{1}\right] + \left[\frac{1}{jw}\right] - \left[\frac{1}{1+jw}\right].$$

Let us define $\Im^{-1}$ to be the inverse Fourier transform. Then by using Table 4.1 again we get

$$y(t) = \Im^{-1}\left[\frac{\pi\delta(w)}{1} + \frac{1}{jw}\right] - \Im^{-1}\left[\frac{1}{1+jw}\right].$$

Finally,

$$y(t) = u(t) - e^{-t}u(t).$$

Example 4.13

Let $x_1(t) = e^{-t}u(t)$ and $x_2(t) = \cos(2t)$. Find $y(t) = x_1(t)*x_2(t)$.

SOLUTION

Using the convolution property we write

$$y(t) = x_1(t) * x_2(t) \Leftrightarrow Y(w) = X_1(w)X_2(w).$$

With the values for $X_1(w)$ and $X_2(w)$ substituted using the tables, we have

$$Y(w) = \left[\frac{1}{1+jw}\right]\left[\pi\delta(w-2) + \pi\delta(w+2)\right]$$

$$Y(w) = \left[\frac{\pi\delta(w-2)}{1+jw}\right] + \left[\frac{\pi\delta(w+2)}{1+jw}\right].$$

Using the fact that

$$F(w)\delta(w-w_0) = F(w_0)\delta(w-w_0),$$

we write $Y(w)$ as

$$Y(w) = \left[\frac{\pi\delta(w-2)}{1+j2}\right] + \left[\frac{\pi\delta(w+2)}{1-j2}\right].$$

Notice that $\frac{\pi}{1+j2}$ and $\frac{\pi}{1-j2}$ are constant numbers. Therefore, by utilizing Table 4.1 again we can invert transform $Y(w)$ to write

$$y(t) = \left[\frac{\pi}{1+j2}\right]\frac{1}{2\pi}e^{2jt} + \left[\frac{\pi}{1-j2}\right]\frac{1}{2\pi}e^{-2jt}.$$

In the above equation for $y(t)$, the terms

$$\left[\frac{\pi}{1+j2}\right]\frac{1}{2\pi}e^{2jt}$$

and

$$\left[\frac{\pi}{1-j2}\right]\frac{1}{2\pi}e^{-2jt}$$

are complex conjugate functions. If you add two complex conjugate functions, the sum is twice the real part of the complex number. Using this fact we can write $y(t)$ as

$$y(t) = 2\text{Real}\left\{\left[\frac{\pi}{1+j2}\right]\frac{1}{2\pi}e^{2jt}\right\}$$

$$y(t) = 2\text{Real}\left[\frac{1}{2}\left(\frac{1-2j}{5}\right)\left(\cos(2t) + j\sin(2t)\right)\right],$$

and finally

$$y(t) = [1/10]\left[2\cos(2t) + 4\sin(2t)\right].$$

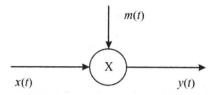

FIGURE 4.3 Sending end of transmission.

Example 4.14

Before we send an information signal through a communication medium, the information signal is converted to another signal via a process called modulation. Let us look at the block diagram in Figure 4.3.

$x(t)$ is the signal that carries the information, $m(t)$ is the signal that we call the carrier, and $y(t)$ is the signal that the receiver gets as input, assuming perfect transmission. Let $m(t) = \cos(w_0 t)$, where w_0 is the carrier frequency. As shown in the multiplier block diagram in Figure 4.3

$$y(t) = x(t)m(t) = x(t)\cos(w_0 t).$$

From Table 4.2 we see that multiplication in the time domain is convolution in the frequency domain. With the help of Table 4.1 we can write the Fourier transform of $y(t)$ as

$$Y(w) = \frac{1}{2\pi}\left[X(w)\right] * \pi\left[\delta(w - w_0) + \delta(w + w_0)\right] = \frac{X(w) * M(w)}{2\pi}$$

with

$$M(w) = \pi\left[\delta(w - w_0) + \delta(w + w_0)\right].$$

$M(w)$ will have the shape shown in Figure 4.4.

Let us look at the band limited signal $x(t)$ in the frequency domain. Assume $X(w)$ has the form as shown in Figure 4.5.

The convolution of $X(w)$ and $M(w)$ is presented by the graph in Figure 4.6, where $|X(w)|$ is shifted twice (due to the presence of the two impulses) to the left and to the right by w_0. But at the output $y(t)$, we are interested in our transmitted signal $x(t)$.

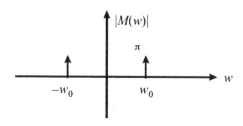

FIGURE 4.4 $m(t)$ magnitude spectra.

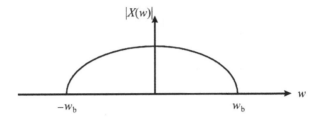

FIGURE 4.5 $x(t)$ magnitude spectra.

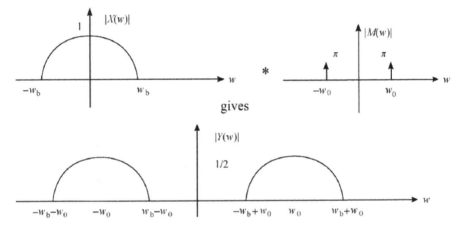

FIGURE 4.6   The convolution of $x(t)$ and $m(t)$ in the frequency domain.

Let us introduce another signal and call it $s(t)$ as seen in Figure 4.7. From Figure 4.7 we can see that

$$s(t) = y(t)\cos(w_0 t).$$

Taking the Fourier transform of this equation for $s(t)$ we will have

$$s(t) = y(t)\cos(w_0 t) \Leftrightarrow S(w) = \frac{1}{2\pi}[Y(w)] * \pi[\delta(w - w_0) + \delta(w + w_0)].$$

This process now is called demodulation and it takes place at the receiving end of the transmission media.

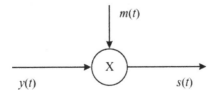

FIGURE 4.7   The end of transmission.

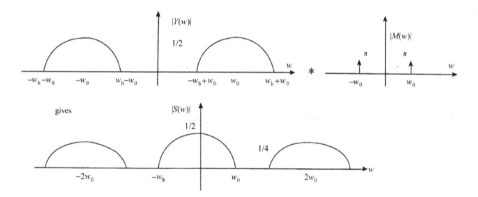

FIGURE 4.8   Convolution of $y(t)$ and $m(t)$ in the frequency domain.

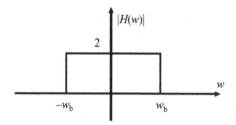

FIGURE 4.9   Low-pass filter.

Let us look now graphically at the convolution process as seen in Figure 4.8. To get our desired signal $x(t)$ at the receiving end of the transmission media, we pass $s(t)$ through a low-pass filter with $H(w)$ as shown in Figure 4.9.

4.4   ENERGY OF NONPERIODIC SIGNALS

The energy of the signal $x(t)$ is defined as

$$E = \int_{-\infty}^{\infty} |x(t)|^2 \, dt = \int_{-\infty}^{\infty} x(t) x^*(t) \, dt,$$

where * means conjugate.

Taking the Fourier transform of $x^*(t)$ we get

$$E = \int_{-\infty}^{\infty} x(t) \left[\frac{1}{2\pi} \int_{-\infty}^{\infty} X^*(w) e^{-jwt} \, dw \right] dt$$

$$E = \frac{1}{2\pi} \int_{-\infty}^{\infty} X^*(w) \left[\int_{-\infty}^{\infty} x(t) e^{-jwt} \, dt \right] dw.$$

The last integral term in the above equation is the Fourier transform of $x(t)$, $X(w)$. Therefore,

$$E = \frac{1}{2\pi} \int_{-\infty}^{\infty} X^*(w) X(w) dw = \frac{1}{2\pi} \int_{-\infty}^{\infty} |X(w)|^2 dw,$$

that is, what we call the theorem of Parseval. Since $|X(w)|$ is an even function, we have

$$E = \frac{1}{\pi} \int_{0}^{\infty} |X(w)|^2 dw. \tag{4.8}$$

4.5 THE ENERGY SPECTRAL DENSITY OF A LINEAR SYSTEM

Consider a linear time-invariant system with input $x(t)$, impulse response function $h(t)$, and output $y(t)$. In real time we write the output as

$$y(t) = x(t) * h(t),$$

where * means convolution. In the Fourier transform language we write the output $y(t)$ as

$$Y(w) = X(w) H(w),$$

and by taking magnitudes we get

$$|Y(w)|^2 = |X(w) H(w)|^2,$$

which simplifies to

$$|Y(w)|^2 = |X(w)|^2 |H(w)|^2. \tag{4.9}$$

Therefore, the energy spectral density for the output of the system is the product of the energy spectral density of the input times the squared magnitude of the system's transfer function $H(w)$.

4.6 SOME INSIGHTS: NOTES AND A USEFUL FORMULA

Fourier series are limited as to their use. They only represent signals of periodic nature. Many important functions like the isolated pulse and the decaying exponential are nonperiodic and therefore cannot be represented by the Fourier series. Using the Fourier transform we were able to represent such functions.

Fourier series coefficients c_n in the previous chapter were used to study and examine the frequency components of the signal under investigation. The Fourier transform allows us to look at the Fourier transform of the impulse function $h(t)$, and thus, study the frequency response of the system $h(t)$ represents. From $H(jw)$ we can deduce the magnitude and the phase of the system. If the system under investigation is to pass a signal that has many unknown frequencies, plotting $H(jw)$ vs. w will tell us what frequencies the system will pass and what frequencies it will block or attenuate.

Knowing the transfer function $H(jw)$, we can find its magnitude and phase at a particular frequency. If the system $H(jw)$ represents is subject to a sinusoidal input signal at that particular frequency of $H(jw)$, the steady-state output $y_{ss}(t)$ of the system can be evaluated as

$$y_{ss}(t) = |x(t)||H(jw)|\cos(wt + \theta + \varphi), \tag{4.10}$$

where

$$x(t) = A\cos(wt + \theta)$$

and φ is the phase of $H(jw)$ at a particular given w.
If

$$x(t) = A\sin(wt + \theta),$$

then

$$y_{ss}(t) = |x(t)||H(jw)|\sin(wt + \theta + \varphi). \tag{4.11}$$

This says that the output of a linear time-invariant system, if subject to a sinusoidal input, will have a steady-state solution equal to the magnitude of the input signal multiplied by the magnitude of the transfer function of the system, evaluated at the frequency of the input signal and shifted by the phase angle of the transfer function evaluated at the input frequency as well. Some notes on Fourier transform are worth mentioning:

1. The magnitude spectrum of the Fourier transform of a signal $x(t)$ is even and the phase spectrum is odd.
2. Shifting the signal $x(t)$ in time does not alter its magnitude.
3. Time compression of a signal $x(t)$ corresponds to frequency expansion of $X(w)$.
4. Time expansion of a signal $x(t)$ corresponds to frequency compression of $X(w)$.
5. A time-limited signal $x(t)$ has a Fourier transform, $X(w)$, which is not band limited (by band limited we mean the frequency band).
6. If $X(w)$ is band limited, then $x(t)$ is not time limited.

4.7   END-OF-CHAPTER EXAMPLES

EOCE 4.1

Let the Fourier transforms of $x_1(t)$ and $x_2(t)$ be $X_1(w)$ and $X_2(w)$, respectively. Let $\mathfrak{I}$ denote "the Fourier transform of." Is $\mathfrak{I}(ax_1(t) + bx_2(t)) = a\,\mathfrak{I}(x_1(t)) + b\,\mathfrak{I}(x_2(t))$? Assume a and b are constants.

SOLUTION

Using the defining integral of the Fourier transform and properties of integrals we have

$$\mathfrak{I}\big(ax_1(t) + bx_2(t)\big) = \int_{-\infty}^{\infty} \big(ax_1(t) + bx_2(t)\big)e^{-jwt}dt$$

$$\mathfrak{I}\big(ax_1(t) + bx_2(t)\big) = a\int_{-\infty}^{\infty} x_1(t)e^{-jwt}dt + b\int_{-\infty}^{\infty} x_2(t)e^{-jwt}dt$$

$$\mathfrak{I}\big(ax_1(t) + bx_2(t)\big) = aX(w) + bX(w) = a\mathfrak{I}\big(x_1(t)\big) + b\mathfrak{I}\big(x_2(t)\big).$$

This is what we call the linearity property of the Fourier transform.

EOCE 4.2

If $X(w)$ is the Fourier transform of $x(t)$, what is the relation between $X(t)$ and $2\pi x(-w)$?

SOLUTION

From the defining integral

$$X(w) = \int_{-\infty}^{\infty} x(t)e^{-jwt}dt,$$

and

$$x(t) = \frac{1}{2\pi}\int_{-\infty}^{\infty} X(w)^{jwt}\,dw.$$

If we set $t = -t$, then the last integral becomes

$$x(-t) = \frac{1}{2\pi}\int_{-\infty}^{\infty} X(w)e^{-jwt}dw$$

or

$$2\pi x(-t) = \int_{-\infty}^{\infty} X(w)e^{-jwt}dw.$$

Now replace w by t to get

$$2\pi x(-w) = \int_{-\infty}^{\infty} X(t)e^{-jwt}dt.$$

This last equation says that the Fourier transform of $X(t)$ is $2\pi x(-w)$, where $X(t)$ is $X(w)$ evaluated at $w = t$.

EOCE 4.3

If a is a positive constant, what is the Fourier transform of $x(at)$?

SOLUTION

Let us look at the familiar integral

$$\int_{-\infty}^{\infty} x(at)e^{-jwt}dt.$$

If we let $at = m$, then $dt = dm/a$. The integral becomes

$$1/a \int_{-\infty}^{\infty} x(m)e^{-j(m/a)w}dm = \frac{1}{a}X(w/a).$$

Therefore,

$$\Im(x(at)) = \frac{1}{a}X(w/a).$$

EOCE 4.4

What is the Fourier transform of the shifted signal $x(t - a)$?

SOLUTION

Using the defining integral, we have

$$\Im(x(t-a)) = \int_{-\infty}^{\infty} x(t-a)e^{-jwt}dt.$$

Let $t - a = m$, then $dt = dm$. The above integral becomes

$$\Im(x(t-a)) = \int_{-\infty}^{\infty} x(m)e^{-jw(m+a)}dm$$

$$\Im(x(t-a)) = e^{-jwa}\int_{-\infty}^{\infty} x(m)e^{-jwm}dm.$$

Now it is clear from the last step that

$$\Im(x(t-a)) = e^{-jwa}X(w).$$

EOCE 4.5

What is the Fourier transform of $e^{jat}x(t)$?

SOLUTION

By using the defining integral again we can write

$$\Im(e^{jat}x(t)) = \int_{-\infty}^{\infty} e^{jat}x(t)e^{jwt}dt$$

$$\Im(e^{jat}x(t)) = \int_{-\infty}^{\infty} x(t)e^{-j(w-a)t}dt = X(w-a).$$

EOCE 4.6

What is the Fourier transform of $x(t)\cos(at)$?

SOLUTION

$\cos(at)$ can be written in exponential form as

$$\cos(at) = \frac{e^{jat} + e^{-jat}}{2}.$$

Taking the Fourier transform we get

$$\Im(x(t)\cos(at)) = \Im\left(x(t)\frac{e^{jat} + e^{-jat}}{2}\right)$$

$$= \Im\left(x(t)\frac{e^{jat}}{2}\right) + \Im\left(x(t)\frac{e^{-jat}}{2}\right).$$

By simplification we arrive at

$$\Im\big(x(t)\cos(at)\big) = \frac{1}{2}\Im\big(x(t)e^{jat}\big) + \frac{1}{2}\Im\big(x(t)e^{-jat}\big)$$

$$= \frac{1}{2}X(w-a) + \frac{1}{2}X(w+a).$$

EOCE 4.7

What is the Fourier transform of the derivative of x(t)?

SOLUTION

$$\Im\left(\frac{d}{dt}x(t)\right) = \int_{-\infty}^{\infty}\left(\frac{d}{dt}x(t)\right)e^{-jwt}dt$$

If we let $u = e^{-jwt}$ and $\left(\dfrac{d}{dt}x(t)\right)dt$, then

$$\Im\left(\frac{d}{dt}x(t)\right) = e^{-jwt}x(t)\Big|_{-\infty}^{\infty} + jw\int_{-\infty}^{\infty}x(t)e^{-jwt}dt.$$

The above equation was obtained by integrating by parts.

$$\Im\left(\frac{d}{dt}x(t)\right) = jw\int_{-\infty}^{\infty}x(t)e^{-jwt}dt$$

$$\Im\left(\frac{d}{dt}x(t)\right) = jwX(w).$$

EOCE 4.8

What is the Fourier transform of the second and the nth derivative of x(t)?

SOLUTION

By using the properties of the Fourier transform we write

$$\Im\left(\frac{d^2}{dt}x(t)\right) = jw\Im\left(\frac{d}{dt}x(t)\right) = jw(jw)X(w)$$

$$\Im\left(\frac{d^2}{dt}x(t)\right) = (jw)^2 X(w) = -w^2 X(w).$$

Similarly,

$$\Im\left(\frac{d^n}{dt}x(t)\right) = (jw)^n X(w).$$

EOCE 4.9

Consider the signal in Figure 4.10 and its derivative in Figure 4.11. What is the Fourier transform of $x(t)$?

SOLUTION

Let us start by taking the Fourier transform of $\frac{d}{dt}x(t)$. We can write $x(t)$ mathematically as

$$x(t) = \begin{cases} 0 & t < 0 \\ t & 0 \le t < 1 \\ -t+2 & 1 \le t < 2 \\ 0 & t \ge 2 \end{cases}.$$

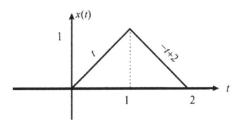

FIGURE 4.10   Signal for EOCE 4.9.

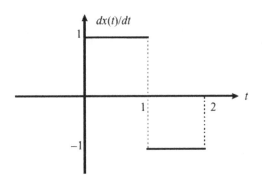

FIGURE 4.11   Signal for EOCE 4.9.

Its derivative can also be written mathematically as

$$\frac{d}{dt}x(t) = \begin{cases} 1 & 0 \le t < 1 \\ -1 & 1 \le t < 2 \\ 0 & \text{otherwise} \end{cases}.$$

Taking the Fourier transform of $\frac{d}{dt}x(t)$ piece by piece we get

$$\Im\left(\frac{d}{dt}x(t)\right) = \begin{cases} \displaystyle\int_0^1 e^{-jwt}dt = \frac{1}{-jw}\left[e^{jw} - 1\right] \\[4mm] \displaystyle\int_0^2 e^{-jwt}dt = \frac{1}{-jw}\left[e^{-2jw} - e^{-jw}\right] \end{cases}.$$

Therefore, the Fourier transform of $\frac{d}{dt}x(t)$ is

$$\Im\left(\frac{d}{dt}x(t)\right) = \frac{1}{jw}\left[e^{-2jw} - 2e^{-jw} + 1\right].$$

Now we can use entry 6, Table 4.2, to find the Fourier transform of x(t) as

$$\Im(x(t)) = X(w) = \Im(dx(t)/dt)/jw$$

$$\Im(x(t)) = X(w) = \frac{e^{-2jw} - 2e^{-jw} + 1}{(jw)(jw)} = \frac{-e^{-2jw} + 2e^{-jw} - 1}{w^2}.$$

Note that in many situations, taking the Fourier transform of the derivative of a signal to calculate the Fourier transform of the signal itself is easier when the signal is hard to integrate using the defining integral directly.

EOCE 4.10

We have seen in this chapter that convolution in the time domain is multiplication in the frequency domain. If

$$\Im(a(t) * b(t)) = \int_0^t a(\tau)b(t - \tau)d\tau = A(w)B(w),$$

what is x(t) if

$$X(w) = \frac{1}{5 + jw}\frac{1}{6 + jw}?$$

SOLUTION

We can use partial fraction expansion to find $x(t)$, but instead we will do the following.

Let

$$A(w) = \frac{1}{5 + jw}$$

and

$$B(w) = \frac{1}{6 + jw}.$$

Therefore,

$$a(t) = e^{-5t}u(t).$$

and

$$b(t) = e^{-6t}u(t).$$

But according to what is given in this example,

$$x(t) = \int_0^t a(\tau)b(t-\tau)d\tau$$

$$x(t) = \int_0^t e^{-5\tau}e^{-6(t-\tau)}d\tau = e^{-6t}\int_0^t e^{\tau}d\tau,$$

and finally

$$x(t) = \left(e^{-5t} - e^{-6t}\right)u(t).$$

EOCE 4.11

Consider the following differential equation

$$\frac{d^2}{dt}y - 5\frac{d}{dt} + 6y = x(t),$$

where $x(t)$ is the input and $y(t)$ is the output.

If $h(t)$ is the impulse response (the transfer function of the system as seen in Chapter 3), $H(w)$ is the Fourier transform of $h(t)$. $H(w)$ is the ratio of $Y(w)$ to $X(w)$. For a particular input $x(t)$,

$$\Im\big(y(t)\big)=\Im\big((x(t)*h(t)\big)=\Im\left(\int_{0}^{t}x(\tau)h(t-\tau)d\tau\right)=X(w)H(w).$$

We can see that if $x(t)$ is given, we can use the Fourier transform to find $y(t)$. Let $x(t)=e^{-2t}$ for $t>0$. Find $y(t)$ using the Fourier techniques.

SOLUTION

We start by finding $H(w)$. We do that by taking the Fourier transform of the differential equation with the input as $x(t)$ to get

$$\left(-w^{2}\right)Y(w)-5jwY(w)+6Y(w)=X(w).$$

Taking $Y(w)$ common we get

$$Y(w)\left[-w^{2}-5jw+6\right]=X(w)$$

$$H(w)=Y(w)/X(w)$$

$$H(w)=\frac{1}{-w^{2}-5jw+6}=\frac{1}{(jw)^{2}-5jw+6}=\frac{1}{jw-3}-\frac{1}{jw-2}.$$

But

$$Y(w)=H(w)X(w).$$

Therefore,

$$Y(w)=\left[\frac{1}{jw-3}-\frac{1}{jw-2}\right]\left[\frac{1}{jw+2}\right].$$

Multiply out to get

$$Y(w)=\left[\frac{1}{jw-3}\frac{1}{jw+2}-\frac{1}{jw-2}\frac{1}{jw+2}\right].$$

$$Y(w)=\frac{1/5}{jw-3}-\frac{1/5}{jw+2}-\frac{1/4}{jw-2}+\frac{1/4}{jw+2},$$

and $y(t)$ is

$$y(t)=1/5\left[e^{3t}-e^{-2t}\right]u(t)+1/4\left[-e^{2t}-e^{-2t}\right]u(t).$$

EOCE 4.12

Let $y(t)$ be the output of a linear system with $Y(w)$ as its Fourier transform. If $Y(w)$ can be written in polar form as

$$Y(w) = M(w)e^{j\theta(w)},$$

where $M(w)$ and $\theta(w)$ are the magnitude and the angle of $Y(w)$, respectively, is the energy equation

$$\int_{-\infty}^{\infty} y^2(t)\,dt = \frac{1}{2\pi}\int_{-\infty}^{\infty} M^2(w)\,dw$$

true? What is this relation?

SOLUTION

Let us look at the defining integral of the Fourier transform

$$Y(w) = \int_{-\infty}^{\infty} y(t)e^{-jwt}\,dt.$$

$Y(w)$ conjugated and reflected is

$$\overset{*}{Y}(-w) = Y(w) = \int_{-\infty}^{\infty} y(t)e^{-jwt}\,dt,$$

where * represents complex conjugate. Similarly,

$$Y(w-\tau) = \overset{*}{Y}(\tau - w).$$

If we convolve $Y(w)$ with $Y(w)$, we will have

$$Y(w)\overset{*}{Y}(w) = \int_{-\infty}^{\infty} Y(\tau)Y(w-\tau)\,d\tau = \int_{-\infty}^{\infty} Y(\tau)\overset{*}{Y}(\tau - w)\,d\tau.$$

We know that

$$\Im\left(y^2(t)\right) = \frac{1}{2\pi}Y(w)\overset{*}{Y}(w) = \frac{1}{2\pi}\int_{-\infty}^{\infty} Y(\tau)\overset{*}{Y}(\tau - w)\,d\tau.$$

Also

$$\Im\left(y^2(t)\right) = \int_{-\infty}^{\infty} y^2(t)e^{-jwt}\,dt.$$

If we let $w = 0$ in the above two equations, then

$$\Im\left(y^2(t)\right) = \frac{1}{2\pi} \int\limits_{-\infty}^{\infty} Y(\tau) \dot{Y}(\tau) d\tau,$$

and

$$\Im\left(y^2(t)\right) = \int\limits_{-\infty}^{\infty} y^2(t) dt.$$

Now let $w = \tau$ to get

$$\Im\left(y^2(t)\right) = \frac{1}{2\pi} \int\limits_{-\infty}^{\infty} Y(w) \dot{Y}(w) dw = \frac{1}{2\pi} \int\limits_{-\infty}^{\infty} M^2(w) dw.$$

But

$$\Im\left(y^2(t)\right) = \int\limits_{-\infty}^{\infty} y^2(t) dt.$$

Therefore,

$$\int\limits_{-\infty}^{\infty} y^2(t) dt = \frac{1}{2\pi} \int\limits_{-\infty}^{\infty} M^2(w) dw,$$

which is the Parseval theorem.

$M^2(w)$ is the energy spectrum, and the integral from $-\infty$ to $+\infty$ is the energy in the signal.

EOCE 4.13

If $y(t) = e^{-t}u(t)$, use the Parseval theorem to find the total energy in $y(t)$.

SOLUTION

We will use the time domain and the frequency domain to find the energy in the signal $y(t)$.

In the time domain the energy is given as

$$\int\limits_{-\infty}^{\infty} y^2(t) dt = \int\limits_{0}^{\infty} e^{-2t} dt = -\frac{1}{2} e^{-2t} \Big|_{0}^{\infty} = \frac{1}{2}.$$

In the frequency domain the energy is given as

$$\frac{1}{2\pi}\int\limits_{-\infty}^{\infty}M^2(w)\,dw$$

with $M(w)$ the magnitude of $Y(w)$.

$$Y(w)=\frac{1}{jw+1}$$

and

$$M(w)=\sqrt{\frac{1}{w^2+1}}.$$

Therefore, the energy is given as

$$\frac{1}{2\pi}\int\limits_{-\infty}^{\infty}M^2(w)\,dw=\frac{1}{2\pi}\int\limits_{-\infty}^{\infty}\frac{1}{w^2+1}\,dw$$

or using the symmetry property, we get

$$\frac{1}{\pi}\int\limits_{0}^{\infty}M^2(w)\,dw=\frac{1}{\pi}\tan^{-1}(w)\Big|_{0}^{\infty}=\frac{1}{2}.$$

EOCE 4.14

Consider the circuit in Figure 4.12 with the input as $x(t)$ and the output as the voltage across the capacitor, $y(t)$. If $x(t) = \cos(10t)$, what is the capacitor steady-state voltage?

SOLUTION

The differential equation describing the circuit is

$$\frac{d}{dt}yt+y(t)/RC=x(t)/RC.$$

FIGURE 4.12 Circuit for EOCE 4.14.

Taking the Fourier transform of the differential equation above we get

$$jwY(w) + Y(w)/RC = X(w)/RC,$$

with $x(t) = \cos(10t)$

$$X(w) = \frac{\pi}{2}\left[\delta(w+10) + \delta(w+10)\right].$$

Therefore,

$$Y(w)\left[jw + 1/RC\right] = \frac{\frac{\pi}{2}\left[\delta(w-10) + \delta(w+10)\right]}{RC}.$$

Simplifying the above equation we get

$$Y(w) = \frac{\frac{\pi}{2}\left[\delta(w-10) + \delta(w+10)\right]}{(jw+1/RC)RC}.$$

$$Y(w) = \frac{\frac{\pi}{2}\left[\delta(w-10)\right]}{(10j+1/RC)RC} + \frac{\frac{\pi}{2}\left[\delta(w+10)\right]}{(-j10+1/RC)RC},$$

which can be written as

$$Y(w) = \frac{\frac{\pi}{2}\left[\delta(w-10)\right]}{(RC10j+1)} + \frac{\frac{\pi}{2}\left[\delta(w+10)\right]}{(-RCj10+1)},$$

Using the Fourier transform table, we can write $y(t)$ as

$$y(t) = \frac{\pi}{2\pi 2(RCj10+1)}e^{j10t} + \frac{\pi}{2\pi 2(-RCj10+1)}e^{-j10t}$$

or

$$y(t) = \frac{1/4}{(RCj10+1)}e^{j10t} + \frac{1/4}{(-RCj10+1)}e^{-j10t}.$$

Let $(1 + RCj10)$ be written in polar form as $Me^{j\theta}$, then

$$\left(1 - RCj10\right) = Me^{-j\theta}.$$

The output $y(t)$ becomes

$$y(t) = \frac{1}{4M}e^{-j\theta}e^{j10t} + \frac{1}{4M}e^{j\theta}e^{-j10t}$$

$$y(t) = \frac{1}{4M}\left[e^{j(10t-\theta)} + e^{-j(10t-\theta)}\right].$$

Finally,

$$y(t) = \frac{1}{2M}\cos(10t - \theta).$$

EOCE 4.15

What is the Fourier transform of $u(t) = \begin{cases} 1 & t>0 \\ 0 & t<0 \end{cases}$?

SOLUTION

We have seen that $u(t)$ can be written as

$$u(t) = 1/2 + \text{sgn}(t)/2,$$

where $sgn(t)$ is as described in Chapter 1.

$$\Im(u(t)) = \Im(1/2) + \frac{1}{2}\Im(sgn(t)) = \pi\delta(w) + \frac{1}{jw}.$$

EOCE 4.16

What is $x(t)$ if its Fourier transform is $X(w) = sgn(w)$, where

$$\text{sgn}(w) = \begin{cases} -1 & w<0 \\ 1 & w>0 \end{cases} ?$$

SOLUTION

From EOCE 4.2, we saw that the Fourier transform of $X(t)$ is $2\pi x(-w)$, where $X(t)$ is $X(w)$ evaluated at $w = t$.

$$\Im(X(t)) = \Im(sgn(t)) = \frac{2}{jw}$$

Therefore,

$$2/(jw) = 2\pi x(-w).$$

We can solve for $x(-w)$ to get

$$x(-w) = \frac{2}{2jw\pi}.$$

By letting $-w = t$ we get

$$x(t) = \frac{1}{-jt\pi}.$$

4.8 END-OF-CHAPTER PROBLEMS

EOCP 4.1

Find the Fourier transform for each of the following signals.

1. $2u(t - 10) + 10\delta(t - 1)$
2. $\sin(2(t - 1))u(t - 1)$
3. $\cos(3(t - 5))u(t - 5)$
4. $e^{-a(t - b)} u(t - b) + u(t - 2)$
5. $(t - 1)e^{-a(t - 1)} u(t - 1)$
6. $10e^{-2(t-2)^2} u(t - 2)$
7. $2\delta(t - 2) + 3\delta(t - 4)$
8. $e^{-3t}u(t) + \dfrac{t^5}{5!}e^{-10t}u(t) + u(t - 5)$

EOCP 4.2

Given the following signals, find the Fourier transform of the derivative for each.

1. $tu(t)$
2. $t\sin(t)u(t)$
3. $(e^{-3t} + \sin(t))u(t)$
4. $(\cos wt + \sin wt)u(t) + 10u(t)$
5. $e^{jwt} u(t)$

EOCP 4.3

Find the Fourier transform for each of the following signals, where * indicates convolution.

1. $e^{-t} u(t) * e^{-t} u(t)$
2. $[\sin(t)u(t)*u(t)]*u(t)$
3. $[e^{-t} u(t) + e^{-2t} u(t)]*\sin(t)u(t)$
4. $[e^{-3t}u(t)*u(t)] + u(t)*u(t)$
5. $\delta(t)*\sin(10t)u(t) - \delta(t - 1)*u(t)$

EOCP 4.4

Consider the following systems represented by the differential equations. $x(t)$ is the input, and $y(t)$ is the output. Find a particular solution to each system.

1. $\dfrac{d}{dt}y + 2y = \sin(t)u(t)$

2. $\dfrac{d^2}{dt}y + 3\dfrac{d}{dt}y + 2y = u(t)$

3. $\dfrac{d^2}{dt}y + 3\dfrac{d}{dt}y = e^{-t}u(t)$

4. $\dfrac{d^2}{dt}y + 5y = e^{-t}u(t)$

5. $\dfrac{d^2}{dt}y + \dfrac{d}{dt}y + y = \delta(t)$

EOCP 4.5

Given are the Fourier transforms for some signals. What are the time-domain signals?

1. $\dfrac{1}{jw + 2} + \dfrac{1}{jw + 3}$

2. $\dfrac{1}{-w^2 + 7jw + 10}$

3. $\dfrac{1}{-w^2 + 5jw + 6}$

4. $\dfrac{1}{-w^2 + 4jw + 4}$

5. $\dfrac{2}{jw} + 4\delta(w) + \dfrac{2\pi\delta(w)}{5}$

EOCP 4.6

Use the Fourier transform to find $y(t)$, a particular solution, in each of the systems shown in Figures 4.13–4.17.

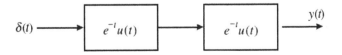

FIGURE 4.13 System for EOCP 4.6.

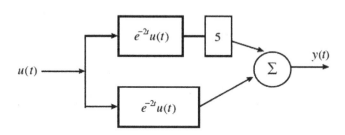

FIGURE 4.14 System for EOCP 4.6.

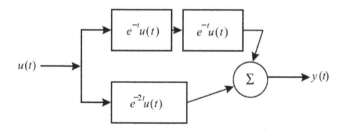

FIGURE 4.15 System for EOCP 4.6.

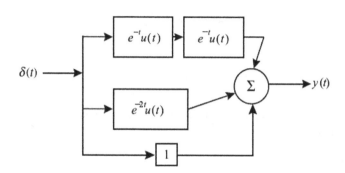

FIGURE 4.16 System for EOCP 4.6.

FIGURE 4.17 System for EOCP 4.6.

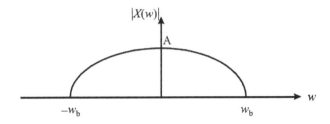

FIGURE 4.18   Signal for EOCP 4.7.

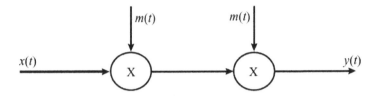

FIGURE 4.19   Block for EOCP 4.7.

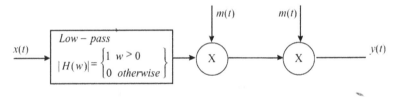

FIGURE 4.20   Block for EOCP 4.7.

EOCP 4.7

A band-limited signal $x(t)$ with its magnitude spectra is shown in Figure 4.18.

In each of the block diagrams shown in Figures 4.19 and 4.20, let $m(t) = \cos(w_0 t)$, and plot the magnitude spectra for $y(t)$.

EOCP 4.8

Consider the following signals. Let $f_o = 60$ Hz.

1. $\cos(w_0 t)$
2. $\sin(w_0 t)$
3. $e^{-3t}u(t)$

Use the Fourier transform to find the energy in each signal in the frequency band between 0 and 10 rad/sec.

EOCP 4.9

What is the Fourier transform for each of the signals in Figures 4.21–4.24?

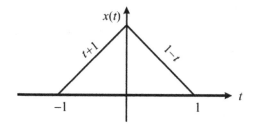

FIGURE 4.21   Signal for EOCP 4.9.

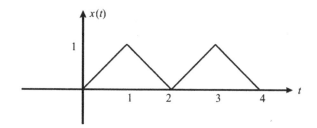

FIGURE 4.22   Signal for EOCP 4.9.

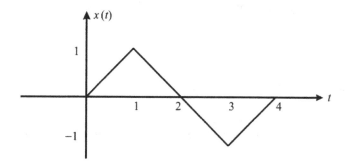

FIGURE 4.23   Signal for EOCP 4.9.

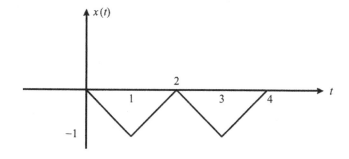

FIGURE 4.24   Signal for EOCP 4.9.

EOCP 4.10

Without using partial fraction expansion, refer to EOCE 4.10 to find $x(t)$ in each of the following if $X(w)$ is

1. $\dfrac{10}{2+jw}\dfrac{1}{4+jw}$

2. $2\pi\delta(w-1)\dfrac{1}{1+jw}$

3. $\dfrac{\pi}{j}[\delta(w-1)-\delta(w+1)]\dfrac{1}{1+jw}$

4. $2\pi\delta(w-10)\dfrac{1}{(1+jw)^2}$

EOCP 4.11

Consider the circuit in Figure 4.25.

1. Find the output voltage $y(t)$, for $t > 0$, using the Fourier transform method.
2. Find the current in the resistor R using the Fourier method.
3. Find the energy in $y(t)$ and the current calculated in 2.

EOCP 4.12

Consider the circuit shown in Figure 4.26.

1. Find $y(t)$ using the Fourier transform method.
2. Find the current in the capacitor using the Fourier transform method.

FIGURE 4.25 Circuit for EOCP 4.11.

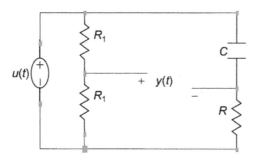

FIGURE 4.26 Circuit for EOCP 4.12.

REFERENCES

Cogdell, J.R. *Foundations of Electrical Engineering*, 2nd ed., Englewood Cliffs, NJ: Prentice-Hall, 1996.

Denbigh, P. *System Analysis and Signal Processing*, Reading, MA: Addison-Wesley, 1998.

Golubitsky, M. and Dellnitz, M. *Linear Algebra and Differential Equations Using MATLAB*, Stamford, CT: Brooks/Cole, 1999.

Harman, T.L., Dabney, J., and Richert, N. *Advanced Engineering Mathematics with MATLAB*, Stamford, CT: Brooks/Cole, 2000.

Moore, H. *MATLAB for Engineers*, Upper Saddle River, NJ: Prentice Hall, 2007.

Nilson, W.J. and Riedel, S.A. *Electrical Circuits*, 6th ed., Englewood Cliffs, NJ: Prentice-Hall, 2000.

Phillips, C.L. and Parr, J.M. *Signals, Systems, and Transforms*, 2nd ed., Englewood Cliffs, NJ: Prentice-Hall, 1999.

Strum, R.D. and Kirk, D.E. *Contemporary Linear Systems*, Boston, MA: PWS, 1996.

The MathWorks. *The Student Edition of MATLAB*, Englewood Cliffs, NJ: Prentice-Hall, 1997.

Wylie, R.C. and Barrett, C.L. *Advanced Engineering Mathematics*, 6th ed., New York, NY: McGraw-Hill, 1995.

Ziemer, R.E., Tranter, W.H., and Fannin, D.R. *Signals Systems Continuous and Discrete*, 4th ed., Englewood Cliffs, NJ: Prentice-Hall, 1998.

5 The Laplace Transform and Linear Systems

5.1 DEFINITION

The Laplace transform, like the Fourier transform, is a frequency-domain representation that makes solution, design, and analysis of linear systems simpler. It also gives some insights about the frequency contents of signals where these insights are hard to see in real-time systems. There are other important uses for the Laplace transform, but we will concentrate only on the issues described in this definition.

5.2 THE BILATERAL LAPLACE TRANSFORM

The bilateral Laplace transform of a signal $x(t)$ is defined as

$$X(s) = \int_{-\infty}^{\infty} x(t)e^{-st}dt, \tag{5.1}$$

where $s = \sigma + jw$.

If $s = jw$ and $\sigma = 0$, the bilateral Laplace transform is then

$$X(s) = \int_{-\infty}^{\infty} x(t)e^{-jwt}dt, \tag{5.2}$$

which is nothing but the Fourier transform of $x(t)$, $X(w)$. If $\sigma \neq 0$ then

$$X(s) = \int_{-\infty}^{\infty} x(t)e^{-(\sigma+jw)t}dt, \tag{5.3}$$

$$X(s) = \int_{-\infty}^{\infty} x(t)e^{-\sigma t}e^{-jwt}dt. \tag{5.4}$$

You can easily see in the above equation that $X(s)$ is, in this case, the Fourier transform of $x(t)e^{-\sigma t}$.

5.3 THE UNILATERAL LAPLACE TRANSFORM

If $x(t) = 0$ for $t < 0$, and $\sigma = 0$, then

$$X(s) = \int_{0^-}^{\infty} x(t)e^{-jwt}dt.$$

We will be interested in the unilateral Laplace transform because, in real life, most systems start at a specific time which we call 0−, where 0− indicates that time starts just before 0.

Again, if $x(t)$ is an input signal to a linear system, then the Laplace transform of $x(t)$ is $X(s)$ and we write the defining integral as

$$X(s) = \int_{0^-}^{\infty} x(t) e^{-st} dt, \tag{5.5}$$

and

$$x(t) \Leftrightarrow X(s)$$

says that the transform goes both ways.

We will present two tables: Table 5.1 for the Laplace transform pairs and Table 5.2 for the properties of the Laplace transform.

TABLE 5.1

Laplace Transform Table: Selected Pairs

$x(t)$	$X(s)$
1. $\delta(t)$	1
2. $\delta(t-a)$	e^{-as}
3. $u(t)$	$\dfrac{1}{s}$
4. $u(t-a)$	$\dfrac{e^{-as}}{s}$
5. $e^{-at} u(t)$	$\dfrac{1}{s+a}$
6. $t^n u(t)\ n = 1, 2, \ldots$	$\dfrac{n!}{s^{n+1}}$
7. $t^n e^{-at} u(t)\ n = 1, 2, \ldots$	$\dfrac{n!}{(s+a)^{n+1}}$
8. $\cos(wt)u(t)$	$\dfrac{s}{s^2 + w^2}$
9. $\sin(wt)u(t)$	$\dfrac{w}{s^2 + w^2}$
10. $e^{-at} \cos(wt)u(t)$	$\dfrac{s+a}{(s+a)^2 + w^2}$
11. $e^{-at} \sin(wt)\ u(t)$	$\dfrac{w}{(s+a)^2 + w^2}$
12. $t \sin(wt)u(t)$	$\dfrac{2ws}{\left(s^2 + w^2\right)^2}$
13. $t \cos(wt)u(t)$	$\dfrac{s^2 - w^2}{\left(s^2 + w^2\right)^2}$

TABLE 5.2
Properties of the Laplace Transform

Time Signal	Laplace Domain
1. $x(t)$	$X(s)$
2. $\displaystyle\sum_{n=1}^{N} a_n x_n(t)$	$\displaystyle\sum_{n=1}^{N} a_n X_{x_n}(s)$
3. $x(t - t_0)$	$X(s)e^{-st_0}$
4. $e^{(at)}x(t)$	$X(s - a)$
5. $x(\alpha t)$	$\dfrac{1}{|\alpha|}X(s/\alpha, \alpha > 0)$
6. $\dfrac{d}{dt}x(t)$	$sX(s) - x(0^-)$
7. $\dfrac{d^2}{dt}x(t)$	$s^2X(s) - sx(0^-) - \dfrac{dx(t)}{dt}(0^-)$
8. $x(t) * h(t)$	$X(s)H(s)$
9. $\displaystyle\int_{0^-}^{\infty} x(\tau)d\tau$	$\dfrac{1}{s}X(s)$
10. $tx(t)$	$\dfrac{-dX(s)}{ds}$
11. $x(t)\sin(w_0 t)$	$\dfrac{1}{2j}\left[X(s - jw_0) - X(s + jw_0)\right]$
12. $x(t)\cos(w_0 t)$	$\dfrac{1}{2}\left[X(s - jw_0) + X(s + jw_0)\right]$
13. $x(o^+)$	$\text{limit}_{s\to\infty}sX(s)$
14. $\text{limit}_{t\to\infty}x(t)$	$\text{limit}_{s\to 0}sX(s)$

5.4 THE INVERSE LAPLACE TRANSFORM

The inverse Laplace transform can be found using Tables 5.1 and 5.2. It can also be found using the following equation

$$x(t) = \frac{1}{2\pi j}\int_{\sigma - jw}^{\sigma + jw} X(s)e^{st}ds, \tag{5.6}$$

where $s = \sigma + jw$.

In the next few examples, we will look at some entries in Table 5.1 and see how they were developed. You need to know that each entry in Table 5.1 is derived using the defining integral

$$X(s) = \int_{-\infty}^{\infty} x(t)e^{-st}dt.$$

Example 5.1

Find the Laplace transform of the impulse signal $\delta(t)$.

SOLUTION

Using the defining integral, we can find the Laplace transform of $\delta(t)$ as

$$X(s) = \int_{0^-}^{\infty} x(t)e^{-st}dt.$$

By substituting for $x(t)$ we get

$$X(s) = \int_{0^-}^{\infty} \delta(t)e^{-st}dt = e^{-s(0)} = 1.$$

Therefore,

$$x(t) = \delta(t) \Leftrightarrow X(s) = 1.$$

Example 5.2

Find the Laplace transform of the signal $u(t)$.

SOLUTION

Using the defining integral, we can find the Laplace transform of $u(t)$ as

$$X(s) = \int_{0^-}^{\infty} x(t)e^{-st}dt.$$

By substituting for $x(t)$ we have

$$X(s) = \int_{0^-}^{\infty} u(t)e^{-st}dt,$$

and finally

$$X(s) = \int_{0}^{\infty} e^{-st}dt = \left.\frac{e^{-st}}{-s}\right|_{0}^{\infty} = \frac{1}{s}.$$

Therefore, we write

$$x(t) = u(t) \Leftrightarrow x(s) = \frac{1}{s}.$$

Example 5.3

Find the Laplace transform of the signal $x(t) = e^{-at} u(t)$.

SOLUTION

We can find the Laplace transform of $e^{-at} u(t)$ using the defining integral and write

$$X(s) = \int_{0^-}^{\infty} x(t) e^{-st} dt$$

$$X(s) = \int_{0^-}^{\infty} e^{-at} u(t) e^{-st} dt$$

$$X(s) = \int_{0}^{\infty} e^{-(a+s)t} dt$$

and finally

$$X(s) = \frac{e^{-(s+a)t}}{-(s+a)} \Bigg|_0^{\infty} = \frac{1}{s+a}.$$

Therefore, we have the pairs

$$x(t) = e^{-at} u(t) \Leftrightarrow X(s) = \frac{1}{s+a}.$$

The other entries in Table 5.1 can be evaluated in the same way.

Example 5.4

Find the inverse Laplace transform of

$$X(s) = 1 - \frac{5}{s+1}.$$

SOLUTION

As Table 5.2 shows, we can use the first entry to inverse transform each term separately. Using Table 5.1 we get

$$\delta(t) \Leftrightarrow 1$$

and

$$\frac{5}{s+1} \Leftrightarrow 5e^{-t}.$$

Therefore,

$$X(s) = 1 - \frac{5}{s+1} \Leftrightarrow x(t) = \delta(t) - 5e^{-t}u(t).$$

Example 5.5

Find the inverse Laplace transform of

$$X(s) = \frac{1}{s^2 + s + 1}.$$

SOLUTION

We can see that the roots of the denominator in the above expression for $X(s)$ are complex. In this case, we can try to avoid using partial fraction expansion to complete the squares in the denominator in $X(s)$ and take advantage of the entries in Table 5.1.

Therefore, we can write

$$X(s) = \frac{1}{\left(s + \frac{1}{2}\right)^2 + \frac{3}{4}} = \frac{1}{\left(s + \frac{1}{2}\right)^2 + \left(\sqrt{\frac{3}{4}}\right)^2}.$$

And by putting the above expression in a form similar to the form in Table 5.1, entry 11, we arrive at

$$X(s) = \frac{1}{\left(\sqrt{\frac{3}{4}}\right)} \cdot \frac{\left(\sqrt{\frac{3}{4}}\right)}{\left(s + \frac{1}{2}\right)^2 + \left(\sqrt{\frac{3}{4}}\right)^2}.$$

We can now look at entry 11 in Table 5.1 and write

$$X(s) = \frac{1}{\left(\sqrt{\frac{3}{4}}\right)} \cdot \frac{\left(\sqrt{\frac{3}{4}}\right)}{\left(s + \frac{1}{2}\right)^2 + \left(\sqrt{\frac{3}{4}}\right)^2} \Leftrightarrow x(t) = \sqrt{\frac{4}{3}} e^{-\frac{1}{2}t} \sin\left(\sqrt{\frac{3}{4}}t\right)u(t).$$

Note that when the roots of the denominator in the s-plane (frequency domain) are complex, use entries 10 and 11 in Table 5.1, depending on the problem at hand.

Example 5.6

Find the inverse Laplace transform of

$$X(s) = \frac{s}{s^2 - 1}.$$

SOLUTION

In this case, the roots are not complex. This $X(s)$ is not shown in Table 5.1. We will use partial fraction expansion as in the following.

$$X(s) = \frac{s}{s^2 - 1} = \frac{A}{s - 1} + \frac{B}{s + 1},$$

where

$$A = (s - 1) \left[\frac{s}{(s - 1)(s + 1)} \right] \Bigg|_{s=1}.$$

Notice here that we can cancel the $(s - 1)$ terms before evaluating at $s = 1$. Therefore, we write

$$A = \left[\frac{s}{(s + 1)} \right] \Bigg|_{s=1} = \frac{1}{2}.$$

B can also be calculated in a similar way as

$$B = (s + 1) \left[\frac{s}{(s - 1)(s + 1)} \right] \Bigg|_{s=-1}$$

or

$$B = \left[\frac{s}{s - 1} \right] \Bigg|_{s=-1} = \frac{1}{2}.$$

Therefore, we now write

$$X(s) = \frac{s}{s^2 - 1} = \frac{A}{s - 1} + \frac{B}{s + 1} = \frac{\frac{1}{2}}{s - 1} + \frac{\frac{1}{2}}{s + 1}.$$

Using Table 5.1 we can inverse transform $X(s)$ and write

$$X(s) = \frac{\frac{1}{2}}{s-1} + \frac{\frac{1}{2}}{s+1} \Leftrightarrow x(t) = \left[\frac{1}{2}e^t + \frac{1}{2}e^{-t}\right]u(t).$$

5.5 BLOCK DIAGRAMS USING THE LAPLACE TRANSFORM

Any transfer function in the s-domain can be written in terms of a fraction: the numerator as a function of s, and the denominator as a function of s, too. We will denote such a transfer function as $H(s)$, the numerator as $N(s)$, and the denominator as $D(s)$. In this case, we can write

$$H(s) = \frac{N(s)}{D(s)}.$$

The transfer function relates the input to an LTI (linear time-invariant) system $x(t)$ to the output of the same system $y(t)$. The transfer function can also be written as

$$H(s) = \frac{Y(s)}{X(s)} = \frac{N(s)}{D(s)}.$$

From the above equation, we can write

$$Y(s) = X(s)H(s).$$

As a block diagram representation, we represent the system in the above equation as in Figure 5.1.

5.5.1 PARALLEL SYSTEMS

Consider three systems represented by their transfer functions $H_1(s)$, $H_2(s)$, and $H_3(s)$, with the single input $X(s)$ and the single output $Y(s)$. These can be represented in a block diagram as shown in Figure 5.2.

$$X(s) \longrightarrow \boxed{H(s)} \longrightarrow Y(s)$$

FIGURE 5.1 General block diagram.

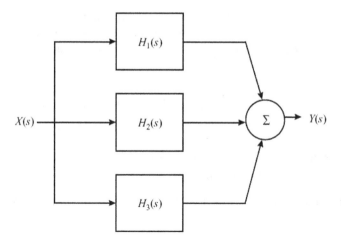

FIGURE 5.2 Parallel system.

In Figure 5.2, the output is

$$Y(s) = (H_1(s) + H_2(s) + H_3(s)) X(s).$$

If we have more transfer functions, the procedure is the same where you add all the individual transfer functions and multiply by the input $X(s)$ to get $Y(s)$. You can find the equivalent transfer function for the whole system as

$$H(s) = \frac{Y(s)}{X(s)} = H_1(s) + H_2(s) + H_3(s).$$

5.5.2 SERIES SYSTEMS

Consider three systems represented by their transfer functions $H_1(s)$, $H_2(s)$, and $H_3(s)$, with the single input $X(s)$ and the single output $Y(s)$. These can be represented in a block diagram as shown in Figure 5.3.

In Figure 5.3, the output is

$$Y(s) = (H_1(s) H_2(s) H_3(s)) X(s).$$

FIGURE 5.3 Series system.

If we have more transfer functions, the procedure is the same where you multiply all the individual transfer functions first and then multiply the result by the input $X(s)$ to get $Y(s)$. You can find the equivalent transfer function for the whole system as

$$H(s) = \frac{Y(s)}{X(s)} = H_1(s)H_2(s)H_3(s).$$

5.6   REPRESENTATION OF TRANSFER FUNCTIONS AS BLOCK DIAGRAMS

Block diagram representation is not unique, as you will see in the following examples which illustrate the procedure.

Example 5.7

Represent the transfer function

$$H(s) = \frac{1}{s^2 - 1}$$

using block diagrams.

SOLUTION

The representation can be accomplished in many ways.

1. With the relation

$$Y(s) = H(s)X(s),$$

the procedure is straightforward, as shown in Figure 5.4.
2. We can write $H(s)$ as the product of two terms

$$H(s) = \frac{1}{s^2 - 1} = \left(\frac{1}{s-1}\right)\left(\frac{1}{s+1}\right).$$

This is the case of multiplying two transfer functions, and the block is shown in Figure 5.5.

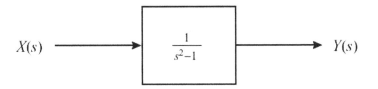

$$X(s) \longrightarrow \boxed{\frac{1}{s^2-1}} \longrightarrow Y(s)$$

FIGURE 5.4   Block for Example 5.7.

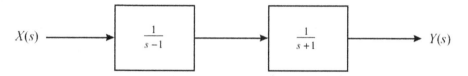

FIGURE 5.5   Block for Example 5.7.

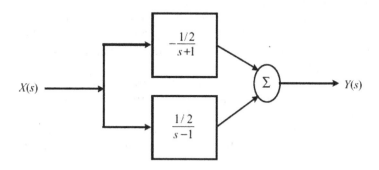

FIGURE 5.6   Block for Example 5.7.

3. We can write $H(s)$ as the sum of two terms

$$H(s) = \frac{1}{s^2 - 1} = \left(\frac{\frac{1}{2}}{s - 1} \right) + \left(\frac{-\frac{1}{2}}{s + 1} \right).$$

This is the case of adding two transfer functions. This block is shown in Figure 5.6.

5.7   PROCEDURE FOR DRAWING THE BLOCK DIAGRAM FROM THE TRANSFER FUNCTION

We will learn this procedure by example.

Consider the following third-order system (we say third order because the denominator polynomial is of degree three) or third-order differential equation represented as the transfer function

$$H(s) = \frac{as^3 + bs^2 + cs + d}{s^3 + es^2 + fs + g},$$

where a, b, c, d, e, f, and g are constants.

It is required that we represent this transfer function in block diagram form.

FIGURE 5.7 Three-integrators block diagram.

The first thing we do here is to make the coefficient of s^3 in the denominator unity. Our transfer function has that characteristic.

Because the order of the denominator in this expression for $H(s)$ is 3, we will need three integrators. Let us precede each integrator by a summer and follow the last integrator by a summer, too. The initial step in the development of drawing this block diagram is shown in Figure 5.7.

Next, we will feedback $Y(s)$ multiplied by the negative of each of the constants e, f, and g, in the denominator of $H(s)$. If we proceed from left to right, the signal $[Y(s)(-g)]$ will be fed to the summer that precedes the first integrator, the signal $[Y(s)(-f)]$ will be fed to the summer that precedes the second integrator, and the signal $[Y(s)(-e)]$ will be fed to the summer that precedes the third integrator. The diagram is shown in Figure 5.8.

Finally, we will feed forward the input signal $[X(s)(d)]$ to the summer that precedes the first integrator, $[X(s)(c)]$ to the summer that precedes the second integrator, $[X(s)(b)]$ to the summer that precedes the third integrator, and $[X(s)(a)]$ to the summer that follows the third integrator. The final block diagram is shown in Figure 5.9.

FIGURE 5.8 Three-integrators block diagram.

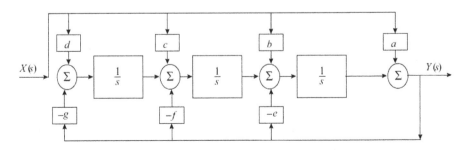

FIGURE 5.9 Three-integrators block diagram.

Notes on the above example:

1. If the order of the numerator in the example at hand is 2, then the feed forward to the summer after the third integrator is zero, and we represent that
   without connection.
2. If the coefficient of s^2 in the numerator in the example at hand is zero, then
   there will be no feed forward signal to the summer that precedes the third
   integrator.
3. If there is no constant term in the numerator (if $d = 0$), then there will be no
   feed forward from $x(t)$ to the first summer before the first integrator.
   We will consider further examples to illustrate these notes.

5.8   SOLVING LTI SYSTEMS USING THE LAPLACE TRANSFORM

Given the system impulse response $h(t)$ and the input $x(t)$, we should be able to find
the output $y(t)$ easily, using the Laplace transform techniques.

Example 5.8

For a certain system, let $h(t) = e^{-t} u(t)$ and $x(t) = \delta(t)$. What is the output of the
system?

SOLUTION

We will take this system into the Laplace domain and find $Y(s)$. We then inverse
transform $Y(s)$ to get $y(t)$. For the impulse response

$$h(t) = e^{-t}u(t) \Leftrightarrow H(s) = \frac{1}{s+1},$$

and for the input

$$x(t) = \delta(t) \Leftrightarrow X(s) = 1.$$

The output of the system in the Laplace domain is

$$Y(s) = X(s)H(s)$$

or

$$Y(s) = (1)\left(\frac{1}{s+1}\right).$$

Using Table 5.1, we can see that the inverse transform of $Y(s)$ is

$$y(t) = e^{-t}u(t).$$

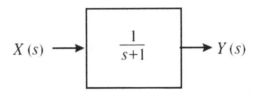

FIGURE 5.10 System for Example 5.8.

Let us look at the system graphically. In the Laplace domain we have the block in Figure 5.10.
    In the time domain, we have the signals as shown in Figure 5.11.

Example 5.9

For a certain system let

$$h(t) = e^{-t}u(t)$$

and

$$x(t) = u(t).$$

What is the output of the system?

SOLUTION

We will take this system into the Laplace domain and find $Y(s)$. We then inverse transform $Y(s)$ to get $y(t)$. For the impulse signal

$$h(t) = e^{-t}u(t) \Leftrightarrow H(s) = \frac{1}{s+1},$$

and for the input

$$x(t) = u(t) \Leftrightarrow X(s) = \frac{1}{s}.$$

FIGURE 5.11 Signals for Example 5.8.

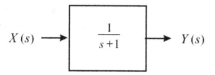

FIGURE 5.12 System for Example 5.9.

FIGURE 5.13 Signals for Example 5.9.

The output is then

$$Y(s) = X(s)H(s)$$

$$Y(s) = \left(\frac{1}{s}\right)\left(\frac{1}{s+1}\right).$$

If we look at Table 5.1, we cannot see an entry similar to Y(s) as given in this example. We will do partial fraction expansion on Y(s) and write

$$Y(s) = \left(\frac{1}{s}\right)\left(\frac{1}{s+1}\right) = \frac{1}{s} - \frac{1}{s+1}.$$

We can now use Table 5.1 and inverse transform each term in Y(s) to get

$$y(t) = u(t) - e^{-t}u(t) = \left(1 - e^{-t}\right)u(t).$$

Graphically, and in the Laplace domain, we have the system shown in Figure 5.12. In the time domain we have the picture in Figure 5.13.

5.9 SOLVING DIFFERENTIAL EQUATIONS USING THE LAPLACE TRANSFORM

We need to remember that the systems we are dealing with here are nothing but dynamical systems that can be represented as differential equations. The differential equation

$$\frac{d}{dt}y(t) + y(t) = x(t)$$

is a first-order differential equation that describes a linear system with $x(t)$ as an input and $y(t)$ as an output. Using Tables 5.1 and 5.2, we can solve differential equations easily.

Example 5.10

Find the output of the following system represented by the differential equation

$$\frac{d}{dt}y(t) + 9y(t) = \delta(t)$$

with the initial condition $y(0^-)$ set to zero.

SOLUTION

In this equation $x(t)$ is given as $\delta(t)$. We need to find $y(t)$. Using Table 5.2, we can write the Laplace equivalent of the differential equation given. We will transform the differential equation term by term and write

$$\left[sY(s) - y(0^-)\right] + 9Y(s) = 1.$$

With the initial condition set to zero, we can solve for $Y(s)$ as

$$Y(s) = \frac{1}{s+9}.$$

We now take the inverse transform of $Y(s)$ using Table 5.1 to get

$$y(t) = e^{-9t}u(t).$$

Example 5.11

Suppose we have the same differential equation as in Example 5.10 with nonzero initial condition $y(0^-) = 1$. What is the output $y(t)$ of this system if $x(t)$ is set to zero?

SOLUTION

We proceed with the solution as we did in the previous example and write the Laplace transform of the system this time as

$$\left[sY(s) - y(0^-)\right] + 9Y(s) = 0$$

$$\left[sY(s) - 1\right] + 9Y(s) = 0.$$

Finally

$$Y(s) = \frac{1}{s+9}.$$

Next, we take the inverse transform to get

$$y(t) = e^{-9t}u(t).$$

Example 5.12

Let us now consider a second-order system described by the differential equation

$$\frac{d^2}{dt^2}y(t) + 2\frac{d}{dt}y(t) = \delta(t).$$

The input is applied at $t = 0$ and the initial conditions are set as

$$y(0^-) = \frac{d}{dt}y(t)\Big|_{t=0^-} = 0.$$

Find the output of this LTI system.

SOLUTION

We can look again at Table 5.2 to transform the given differential equation term by term into its Laplace domain and write

$$\left[s^2Y(s) - \frac{d}{dt}Y(t)\Big|_{t=0^-} - sy(0^-)\right] + 2\left[sY(s) - y(0^-)\right] = 1.$$

Solving for $Y(s)$ we get

$$Y(s) = \frac{1}{s^2 + 2s} = \frac{1}{s(s+2)}.$$

Using partial fractions we write $Y(s)$ as

$$Y(s) = \frac{\frac{1}{2}}{s} + \frac{-\frac{1}{2}}{s+2}.$$

The inverse transform of $Y(s)$, $y(t)$, is obtained using Table 5.1 as

$$y(t) = (1/2)u(t) - (1/2)e^{-2t}u(t),$$

and finally,

$$y(t) = (1/2)\left[1 - e^{-2t}\right]u(t).$$

5.10   THE FINAL VALUE THEOREM

Sometimes we are interested in the final value of the output $y(t)$ as $t \to \infty$. Instead of first solving for $y(t)$ in the time domain, which can be very tedious in some cases, we can use the Laplace transform and write

$$\text{limit}_{t\to\infty}\, y(t) = \text{limit}_{s\to 0}\left[sY(s)\right].$$

To use the theorem above, the roots of the denominator of $sY(s)$ must not be positive.

5.11   THE INITIAL VALUE THEOREM

For the initial value of $y(t)$ we have

$$\text{limit}_{t\to 0^+}\, y(t) = y\left(0^+\right) = \text{limit}_{s\to\infty}\left[sY(s)\right].$$

Example 5.13

If $y(t) = 2e^{-9t}u(t)$, what is $\text{limit}_{t\to\infty} y(t)$?

SOLUTION

In this example, the solution for the output $y(t)$ is given for us. As we said earlier, this solution may not be available, but what we can find in such a case is its transform $Y(s)$ from which we can find

$$\text{limit}_{t\to\infty} y(t) = \text{limit}_{s\to 0}\left[sY(s)\right].$$

In this example, we are lucky and have $y(t)$. We can use this $y(t)$ for checking our answer. Since $u(\infty) = 1$ and $e^{-\infty} = 0$, $\text{limit}_{t\to\infty} y(t) = e^{-9(\infty)} u(\infty) = (0)\,(1) = 0$. According to the final value theorem

$$\text{limit}_{t\to\infty} y(t) = \text{limit}_{s\to 0}\left[sY(s)\right] = \text{limit}_{s\to 0}\left[s\frac{2}{s+9}\right] = \left[\frac{0}{0+9}\right] = 0.$$

Here we see that the two approaches agree.

5.12   SOME INSIGHTS: POLES AND ZEROS

The transfer function $H(s)$ of a LTI system is a very important representation. It tells us about the stability of the system, the poles, the zeros, and the shape of the transients of the output of the system. Using $H(s)$, we can find the steady-state response of the system and the particular solution of the system, all in one shot.

5.12.1 THE POLES OF THE SYSTEM

The poles of the system are the roots of the denominator, the algebraic equation in the variable s, of the transfer function $H(s)$

$$H(s) = \frac{N(s)}{D(s)}.$$

$D(s)$ is a polynomial in s of order equal to the order of the system. The roots of the denominator $D(s)$ are called the poles of the system. These are the same poles we discussed in Chapter 2. Then we called them the eigenvalues of the system. $D(s)$ is actually the characteristic equation of the system, or, as we called it before, the auxiliary equation of the system.

5.12.2 THE ZEROS OF THE SYSTEM

The roots of the numerator $N(s)$ are called the zeros of the system.

5.12.3 THE STABILITY OF THE SYSTEM

The poles of the system determine the stability of the system under investigation. If the real parts of the poles are *all* negative, then the system at hand is stable and the transients will die as time gets larger. The stability of the system is determined by the poles not the zeros of the system. If one of the real parts of the poles is positive, then the system is not stable. You may have positive and negative zeros, but the sign of their values has no effect on the stability of the system.

If $D(s)$ has two roots (second-order system) called α_1 and a_2, then the output will have the general form

$$y(t) = c_1 e^{\alpha_1 t} + c_2 e^{\alpha_2 t},$$

where the constant c's are to be determined. The exponential terms will determine the shape of the transients. If one of the α's is positive, the output will grow without bounds. If the two α's are negative, the output will die as time progresses. The α's are the eigenvalues or the poles of the system.

5.13 END-OF-CHAPTER EXAMPLES

EOCE 5.1

What is the Laplace transform of the derivative of the signal $x(t)$?

SOLUTION

We can use the defining integral

$$X(s) = \int_{0^-}^{\infty} x(t) e^{-st} dt$$

and write

$$L\left(\frac{d}{dt}x(t)\right) = \int_{0^-}^{\infty} \frac{d}{dt}x(t)e^{-st}dt,$$

where L represents "the Laplace transform of."

Let $u = e^{-st}$ and $du = \frac{d}{dt}x(t)dt$, then $du = -se^{-st}dt$ and $v = x(t)$. We now can integrate by parts and write

$$L\left(\frac{d}{dt}x(t)\right) = e^{-st}x(t)\Big|_{0^-}^{\infty} + s\int_{0^-}^{\infty} x(t)e^{-st}dt$$

$$L\left(\frac{d}{dt}x(t)\right) = -x(0^-) + sX(s).$$

Therefore, we can finally write

$$L\left(\frac{d}{dt}x(t)\right) \Leftrightarrow sX(s) - x(0^-).$$

EOCE 5.2

What is the Laplace transform of the second derivative of the signal $x(t)$?

SOLUTION

We can use EOCE 5.1 to solve EOCE 5.2 as in the following.
    The transform of the second derivative is

$$L\left(\frac{d^2}{dt}x(t)\right) = sL\left(\frac{d}{dt}x(t)\right) - \frac{d}{dt}x(0^-).$$

Substituting for $L\left(\frac{d}{dt}x(t)\right)$ from EOCE 5.1 in the above equation, we get

$$L\left(\frac{d^2}{dt}x(t)\right) = s\left[sX(s) - x(0^-)\right] - \frac{d}{dt}x(0^-).$$

By simplification we can finally write

$$L\left(\frac{d^2}{dt}x(t)\right) = s^2X(s) - sx(0^-) - \frac{d}{dt}x(0^-).$$

EOCE 5.3

What is the Laplace transform of $\dfrac{d^3}{dt}x(t)$?

SOLUTION

The transform of the third derivative is

$$L\left(\frac{d^3}{dt}x(t)\right) = sL\left(\frac{d^2}{dt}x(t)\right) - \frac{d^2}{dt}x(0^-).$$

Substituting for $L\left(\dfrac{d^2}{dt}x(t)\right)$ from EOCE 5.2 we get

$$L\left(\frac{d^3}{dt}x(t)\right) = s\left[s^2X(s) - sx(0^-) - \frac{d}{dt}x(0^-)\right] - \frac{d^2}{dt}x(0^-).$$

And with some manipulations, we arrive at

$$L\left(\frac{d^3}{dt}x(t)\right) = s^3X(s) - s^2X(0^-) - s\frac{d}{dt}x(0^-) - \frac{d^2}{dt}x(0^-).$$

EOCE 5.4

Now that we know how to transform a derivative of $x(t)$, what about the integral of $x(t)$?

SOLUTION

Let $X(t)$ be the integral of $x(t)$. Then in this case $\dfrac{d}{dt}X(t) = x(t)$ and

$$L\left(\int_{0^-}^{t} x(u)\,du\right) = L(X(t)).$$

But

$$L(x(t)) = L\left(\frac{d}{dt}X(t)\right) = sL(X(t)) - X(0^-)$$

$$L(x(t)) = sL\left(\int_{0^-}^{t} x(u)\,du - \int_{0^-}^{0^-} x(u)\,du\right)$$

$$L(x(t)) = sL\left(\int_{0^-}^{t} x(u)\,du\right).$$

Therefore

$$L\left(\int_{0^-}^{t}x(u)\,du\right)=\frac{1}{s}L\big(x(t)\big)=\frac{1}{s}X(s).$$

EOCE 5.5

If $x(0^-) = 0$, we claim that the inverse Laplace transform of $sX(s) = \frac{d}{dt}x(t)$. Is this claim true?

SOLUTION

We can use EOCE 5.1 to write

$$L\left(\frac{d}{dt}x(t)\right)=sL\big(x(t)\big)-x\big(0^-\big)-x\big(0^-\big)=sX(s).$$

This implies

$$L^{-1}\big(sX(s)\big)=\frac{d}{dt}x(t).$$

EOCE 5.6

Show that the inverse Laplace transform of $X(s)/s$ is the integral of $x(t)$.

SOLUTION

We know that

$$L\left(\int_{0^-}^{t}x(u)\,du\right)=\frac{1}{s}X(s).$$

This means that

$$L^{-1}\left(\frac{1}{s}X(s)\right)=\int_{0^-}^{t}x(u)\,du.$$

EOCE 5.7

Find the inverse Laplace transform of $\dfrac{1}{s}\dfrac{1}{s+1}$.

SOLUTION

We can use EOCE 5.6 and ignore the term 1/s to find the inverse transform of $1/(s+1)$.

$$\frac{1}{s+1} \Leftrightarrow e^{-t}u(t) = x(t)$$

We can then integrate $x(t)$ and get

$$\int_{0^-}^{t} e^{-u}du = -\left(e^{-t} - 1\right) = -e^{-t} + 1.$$

But

$$\left(-e^{-t} + 1\right)u(t) \Leftrightarrow \frac{1}{s(s+1)},$$

and therefore

$$L^{-1}\left(\frac{1}{s(s+1)}\right) = \left(-e^{-t} + 1\right)u(t).$$

We can use partial fraction expansion to check this answer.

EOCE 5.8

Find the inverse Laplace transform of $\frac{s}{s+1}$ using EOCE 5.5.

SOLUTION

Let us ignore s in the numerator and inverse transform $\frac{1}{s+1}$. By doing that we can write

$$\frac{1}{s+1} \Leftrightarrow e^{-t}u(t).$$

But at $t = 0$, $e^{-t}u(t)$ is not 0. Therefore, we cannot use the EOCE 5.5 result.

EOCE 5.9

Consider the two coupled first-order differential equations with zero initial conditions

$$\frac{d}{dt}y(t) + 2y(t) + z(t) = 0$$

$$\frac{d}{dt}z(t) + z(t) = -u(t).$$

Solve for $y(t)$ using the Laplace transform approach.

SOLUTION

In the Laplace domain the two coupled equations are

$$sY(s) + 2Y(s) + Z(s) = 0 \tag{5.7}$$

$$sZ(s) + Z(s) = -1/s. \tag{5.8}$$

From Equation (5.8) we get $Z(s)$ as

$$Z(s) = -\frac{1}{s(s+1)}.$$

From Equation (5.7) we have

$$Y(s)[s+2] = -Z(s).$$

Substitute for $Z(s)$ to get $Y(s)$ as

$$Y(s) = \frac{1}{s(s+1)(s+2)}.$$

Using partial fraction expansion we get

$$Y(s) = \frac{\frac{1}{2}}{s} + \frac{-1}{s+1} + \frac{\frac{1}{2}}{s+2},$$

and then $y(t)$ is obtained by inverse transforming $Y(s)$ to get

$$y(t) = \frac{1}{2}u(t) - e^{-t}u(t) + \frac{1}{2}e^{-2t}u(t).$$

EOCE 5.10

What is $L(e^{-at}x(t))$?

SOLUTION

Using the defining integral we get

$$L\left(e^{-at}x(t)\right) = \int_{0^-}^{\infty} e^{-at}x(t)e^{-st}dt$$

$$L\left(e^{-at}x(t)\right) = \int_{0^-}^{\infty} x(t)e^{-(s+a)t}dt = X(s)\Big|_{s+a}.$$

Therefore, the pairs are

$$e^{-at}x(t) \Leftrightarrow X(s)\big|_{s+a}.$$

EOCE 5.11

What is the Laplace transform of $e^{-2t}e^{-t}\,u(t)$? Use EOCE 5.10.

SOLUTION

We can use EOCE 5.10 and write

$$e^{-t}u(t) \Leftrightarrow \frac{1}{s+1}$$

and

$$L\left(e^{-2t}e^{-t}u(t)\right) = \frac{1}{s+1}\bigg|_{s+2}$$

$$L\left(e^{-2t}e^{-t}u(t)\right) = \frac{1}{(s+2)+1} = \frac{1}{s+3}$$

or

$$e^{-2t}e^{-t}u(t) \Leftrightarrow \frac{1}{s+3}.$$

EOCE 5.12

Consider a shifted signal $x(t-a)$. What is the Laplace transform of $x(t-a)u(t-a)$?

SOLUTION

Notice that the multiplication $x(t-a)u(t-a)$ starts at $t=a$.
   Using the defining integral we can write

$$L\left(x(t-a)u(t-a)\right) = \int_{0^-}^{\infty} x(t-a)u(t-a)e^{-st}dt$$

$$L\left(x(t-a)u(t-a)\right) = \int_{a}^{\infty} x(t-a)e^{-st}dt.$$

If we let $t-a = m$, then

$$L\left(x(t-a)u(t-a)\right) = \int_{0}^{\infty} x(m)e^{-s(m+a)}dm$$

or

$$L\big(x(t-a)u(t-a)\big) = e^{-sa}\int_0^\infty x(m)e^{-sm}dm = e^{-sa}X(s).$$

EOCE 5.13

Find the Laplace transform of the signal in Figure 5.14.

SOLUTION

Mathematically, the signal can be written as

$$y(t) = (t)\big[u(t) - u(t-1)\big] + (-t+2)\big[u(t-1) - u(t-2)\big].$$

Multiplying the terms out we get

$$y(t) = tu(t) - tu(t-1) + (t-2)u(t-2) + (-t+2)u(t-1).$$

Expanding the terms we will have

$$y(t) = tu(t) - (t+1-1)u(t-1) + (t-2)u(t-2) - (t-1+1)u(t-1) + 2u(t-1)$$

$$y(t) = tu(t) - (t-1)u(t-1) + u(t-1) + (t-2)u(t-2) - (t-1)u(t-1) - u(t-1) + 2u(t-1).$$

We can easily show by using the defining integral that

$$u(t-a) \Leftrightarrow \frac{e^{-sa}}{s}.$$

We can now find $Y(s)$ as

$$Y(s) = \frac{1}{s^2} - \frac{e^{-s}}{s^2} + \frac{e^{-s}}{s} + \frac{e^{-2s}}{s^2} - \frac{e^{-s}}{s^2} - \frac{e^{-s}}{s} + 2\frac{e^{-s}}{s}$$

$$Y(s) = \frac{1}{s^2} - 2\frac{e^{-s}}{s^2} + \frac{e^{-2s}}{s^2} + 2\frac{e^{-s}}{s}.$$

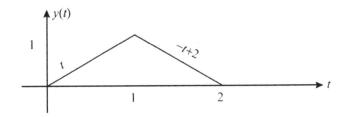

FIGURE 5.14   Signal for EOCE 5.13.

EOCE 5.14

What is $L(tx(t))$?

SOLUTION

Let us differentiate the defining integral

$$X(s) = \int_{0^-}^{\infty} x(t) e^{-st} dt$$

with respect to s on both sides to get

$$\frac{d}{ds} X(s) = \frac{d}{ds} \int_{0^-}^{\infty} x(t) e^{-st} dt$$

$$\frac{d}{ds} X(s) = \int_{0^-}^{\infty} \frac{d}{ds} x(t) e^{-st} dt = - \int_{0^-}^{\infty} tx(t) e^{-st} dt.$$

Therefore

$$-\frac{d}{ds} X(s) = \int_{0^-}^{\infty} tx(t) e^{-st} dt.$$

The last equation simply says that

$$(tx(t)) \Leftrightarrow -\frac{d}{ds} X(s).$$

EOCE 5.15

What is $L(x(t)/t)$ if $\lim_{t \to 0} [x(t)/t]$ exists?

SOLUTION

By definition

$$X(s) = \int_{0^-}^{\infty} x(t) e^{-st} dt.$$

If we integrate both sides from s to ∞ we get

$$\int_s^{\infty} X(s) ds = \int_s^{\infty} \left[\int_{0^-}^{\infty} x(t) e^{-st} dt \right] ds.$$

By interchanging the integrals we get

$$\int_s^\infty X(s)\,ds = \int_{0^-}^{\infty}\int_s^\infty \left[x(t)e^{-st}\,ds\right]dt = \int_{0^-}^{\infty}\left[x(t)\frac{e^{-st}}{-t}\right]_s\,dt$$

or

$$\int_s^\infty X(s)\,ds = \int_{0^-}^{\infty}\left[\frac{x(t)}{t}e^{-st}\right]dt = L\left(\frac{x(t)}{t}\right),$$

and finally

$$L\left(\frac{x(t)}{t}\right) \Leftrightarrow \int_s^\infty X(s)\,ds.$$

EOCE 5.16

What is $L\left(\dfrac{\sin(t)u(t)}{t}\right)$?

SOLUTION

Using EOCE 5.15 we can write

$$L\big(\sin(t)u(t)/t\big) = \int_s^\infty L\big(\sin(t)u(t)\big)\,ds,$$

which simplifies to

$$L\big(\sin(t)u(t)/t\big) = \int_s^\infty \frac{1}{s^2+1}\,ds = \tan^{-1}(s)\Big|_s^\infty,$$

and finally

$$L\big(\sin(t)u(t)/t\big) = \frac{\pi}{2} - \tan^{-1}(s) = \cot^{-1}(s).$$

EOCE 5.17

Consider the following output in the s-domain

$$Y(s) = \frac{1}{s^3+4s^2-2s-2}.$$

What are the initial and the final values of $y(t)$? Is the system stable?

SOLUTION

We will give the solution in the time domain first. We will use MATLAB to do partial fraction expansion and write the script

```
num = [1];
den = [1 4 -2 -2];
[R, P, K] = residue (num, den)
```

The output is

```
R =
0.0499
0.1364
-0.1864
P =
-4.3539
0.8774
-0.5235
K =
[ ]
```

Therefore

$$Y(s) = \frac{0.049}{s + 4.3539} + \frac{0.1364}{s - 0.8774} - \frac{0.1864}{s + 0.5235}.$$

K has no values. It will have values when the order of the numerator is greater than or equal to the order of the denominator.

We can use the Laplace transform table to inverse transform $Y(s)$ to get $y(t)$ as

$$y(t) = 0.049e^{-4.3539t}u(t) + 0.1364e^{0.8774t}u(t) - 0.1864e^{0.5235t}u(t).$$

We notice that this output is unstable because one of the poles of the system is positive. As t approaches ∞, $y(t)$ approaches ∞ and this indicates that we have a positive pole. As t approaches zero, $y(0) = 0.049 + 0.1364 - 0.1864 = -0.001$.

In the s-domain, we can see that one of the poles is positive and that indicates instability. We cannot use the final value theorem to find $y(\infty)$ because one of the poles is positive.

EOCE 5.18

What is the Laplace transform of the signal $x(t)$ that is periodic with period T on the interval $[0\ \infty]$?

SOLUTION

Using the defining integral again we can write

$$L(x(t)) = \int_{0^-}^{\infty} x(t)e^{-st}dt = \int_{0}^{T} x(t)e^{-st}dt + \int_{T}^{2T} x(t)e^{-st}dt + \int_{2T}^{3T} x(t)e^{-st}dt + \dots$$

Let $t = P$ in the first integral, $t = P + T$ in the second integral, $t = P + 2T$ in the third integral, and

so on, to get

$$L(x(t)) = \int_{0^-}^{\infty} x(t)e^{-st}dt = \int_0^T x(p)e^{-st}dp + \int_0^T x(P+T)e^{-s(P+T)}dP$$

$$+ \int_0^T x(P+2T)e^{-st(P+2T)}dP + \ldots$$

Using the fact that $x(P + T) = x(P)$, we get

$$L(x(t)) = \int_0^T x(P)e^{-sP}dP + e^{-sT}\int_0^T x(P+T)e^{-s(P)}dP$$

$$+ e^{-2sT}\int_0^T x(P+2T)e^{-s(P)}dP + \ldots$$

$$L(x(t)) = \int_0^T x(P)e^{-sP}dP + e^{-sT}\int_0^T x(P)e^{-s(P)}dP$$

$$+ e^{-2sT}\int_0^T x(P)e^{-s(P)}dP + \ldots$$

And by taking the common terms we arrive at

$$L(x(t)) = \left[1 + e^{-sT} + e^{-2sT} + \cdots\right]\int_0^T x(P)e^{-s(P)}dP.$$

Also

$$1 + e^{-sT} + e^{-2sT} + e^{-3sT} + \cdots$$

is a geometric series and can be written as the summation

$$\frac{1}{1-e^{-sT}}.$$

Then

$$L(x(t)) = \left[\frac{1}{1+e^{-sT}}\right]\int_0^T x(P)e^{-s(P)}dP.$$

EOCE 5.19

Consider the following system represented by the differential equation

$$\frac{d^2}{dt}y(t) + \frac{d}{dt}y(t) + y(t) = x(t).$$

Assume zero initial conditions and plot the output $y(t)$ vs. time for a step and an impulse input $x(t)$. Is the system stable? What are the final and the initial values of $y(t)$ when the input is step, impulsive, and ramp?

SOLUTION

We can use MATLAB to find $y(t)$. First, we need to find the transfer function that relates the input to the output. To do that we take the Laplace transform of the given differential equation with a general input $x(t)$ to get

$$s^2Y(s) + sY(s) + Y(s) = X(s).$$

Taking $Y(s)$ as the common factor gives

$$Y(s)\left[s^2 + s + 1\right] = X(s),$$

and the transfer function is

$$\frac{Y(s)}{X(s)} = H(s) = \frac{1}{s^2 + s + 1}.$$

At this point, we can find the output $y(t)$ for any input $x(t)$ by inverse transforming

$$Y(s) = H(s)X(s)$$

into its time domain. Notice that $H(s)$ is a ratio of two polynomials. To find the response (output) due to a step input, we type the following MATLAB script.

```
num = [1];
den = [1 1 1];
step (num, den)
```

The output plot is shown in Figure 5.15.
To find the output due to an impulsive input $x(t)$, we type the following MATLAB script.

```
num = [1];
den = [1 1 1 ];
impulse (num, den)
```

The output is shown in Figure 5.16.

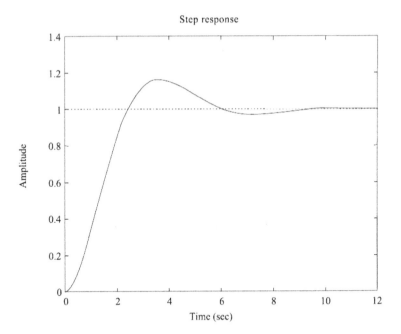

FIGURE 5.15   Plot for EOCE 5.19.

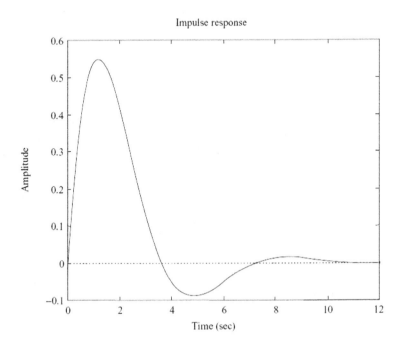

FIGURE 5.16   Plot for EOCE 5.19.

To find the output if the input is $tu(t) = x(t)$, we need to find $Y(s)$ if $x(t) = tu(t)$, because MATLAB does not have a function to calculate a ramp response as we saw for the step and the impulse inputs. In this case, for

$$x(t) = tu(t)$$

we have

$$X(s) = \frac{1}{s^2}.$$

The output $Y(s)$ becomes

$$Y(s) = H(s)X(s) = \frac{1}{s^2}\frac{1}{s^2+s+1}.$$

If we write $Y(s)$ as

$$Y(s) = \frac{1}{s^3+s^2+s}\frac{1}{s},$$

we can use the step response function to calculate the ramp response by writing this script

```
num = [1];
den = [1 1 1 0];
[y, t] = step (num, den);
Plot (y); xlabel ('Time (sec)'), ylabel ('Amplitude')
title ('Response when x(t) = tu(t)');
```

with the plot shown in Figure 5.17.

It can be seen from the graphs that for the step response the initial value is 0 and the final value will approach 1.

For the impulsive input, the initial value is 0 and the final value approaches 0 as well. For the ramp input the initial value is 0 and the final value approaches ∞.

We could use the final and the initial value theorems to confirm the above statements.

For the step input the output is

$$Y(s) = \frac{1}{s}\frac{1}{s^2+s+1}.$$

And using the initial value theorem, we can find the initial value of the output as

$$\text{limit}_{s \to \infty}\left[s\frac{1}{s}\frac{1}{s^2+s+1}\right] = 0 = y(0).$$

We can use the final value theorem to calculate the final value of the output as

$$\text{limit}_{s \to 0}\left[s\frac{1}{s}\frac{1}{s^2+s+1}\right] = 1 = y(\infty).$$

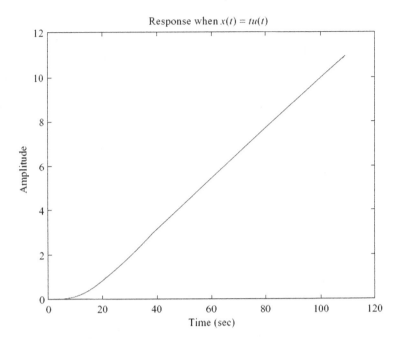

FIGURE 5.17   Plot for EOCE 5.19.

For the impulsive input, the output is

$$Y(s) = \frac{1}{s^2 + s + 1}.$$

The initial value of the output is

$$\lim_{s \to \infty} \left[s \frac{1}{s^2 + s + 1} \right] = 0 = y(0),$$

and the final value is

$$\lim_{s \to 0} \left[s \frac{1}{s^2 + s + 1} \right] = 0 = y(\infty).$$

For the ramp input, the output is

$$Y(s) = \frac{1}{s^2} \frac{1}{s^2 + s + 1}.$$

The initial value of $y(t)$ is

$$\lim_{s \to \infty} \left[s \frac{1}{s^2} \frac{1}{s^2 + s + 1} \right] = 0 = y(0),$$

and its final value is

$$\text{limit}_{s \to 0}\left[s \ \frac{1}{s^2} \ \frac{1}{s^2+s+1}\right] = \infty = y(\infty).$$

To discuss stability of the system, we look at the roots of the denominator of $H(s)$ using MATLAB and type the following command at the MATLAB prompt.

```
poles = roots ([1 1 1])
```

The poles are

```
poles =
-0.5000 + 0.8660i
-0.5000 - 0.8660i
```

We observe that the roots are complex, but the real parts of the roots are negative and this indicates stability of the system represented by $H(s)$.

EOCE 5.20

Consider the following systems:

$$H(s) = \frac{s^3 + s}{s^3 + s^2 + 1}$$

and

$$H(s) = \frac{s^2 + s + 1}{s^3 + 4s^2 + 3s}.$$

1. Find the impulse response of each system.
2. Discuss stability for each system.
3. Draw the block diagram for each system.
4. Find the final value and the initial value for $y(t)$ when the input is an impulse.

SOLUTION

For the first system we write the script EOCE5_201 script with the output as

```
poles_sys1 =
-1.4656
0.2328 + 0.7926i
0.2328 - 0.7926i
```

and the output showing the impulse response is shown in Figure 5.18.

FIGURE 5.18 Plot for EOCE 5.20.

We can see from MATLAB that the real parts of the poles are positive and this indicates instability. It is also seen that the final value of $y(t)$ is ∞. The initial value of $y(t)$ can be calculated from the expression for $Y(s)$,

$$Y(s) = \frac{s^3 + s}{s^3 + s^2 + 1}.$$

The initial value is then

$$\lim_{s \to \infty} \left[s \frac{s^3 + s}{s^3 + s^2 + 1} \right] = \infty = y(\infty),$$

and the final value is

$$\lim_{s \to 0} \left[s \frac{s^3 + s}{s^3 + s^2 + 1} \right] = 0 = y(\infty),$$

which contradicts the results from Figure 5.18. The reason is that one of the poles of $sY(s)$ is positive.

The exact values can be calculated by solving for $y(t)$ in the time domain and then taking the limits to find the initial and final values for $y(t)$.

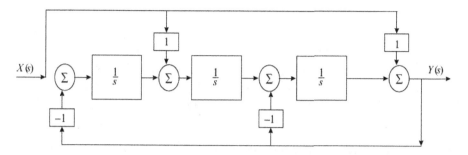

FIGURE 5.19   Block diagram for EOCE 5.20.

The block diagram representing the first system is sketched in Figure 5.19.

The block diagram in Figure 5.19 was drawn with the help of the example presented in this chapter. For the second system we type EOCE5_202 script with the output

```
poles_sys2 =
0
-3
-1
```

and with the impulse response as shown in Figure 5.20.

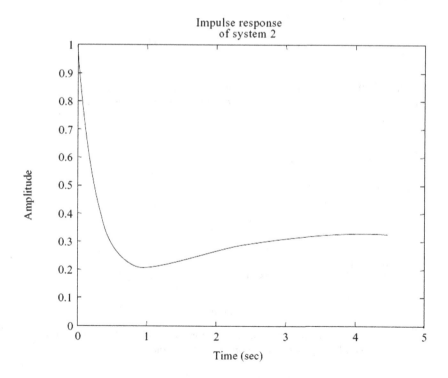

FIGURE 5.20   Plot for EOCE 5.20.

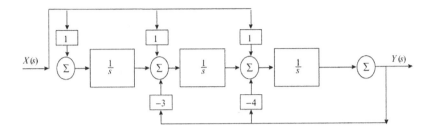

FIGURE 5.21 Block diagram for EOCE 5.20.

We can see from MATLAB that the poles are negative and this indicates stability. The initial and final values of $y(t)$ can be calculated by first looking at the expression

$$Y(s) = \frac{s^2 + s + 1}{s^3 + 4s^2 + 3s}.$$

The initial value for $y(t)$ is

$$\lim_{s \to \infty} \left[s \frac{s^2 + s + 1}{s^3 + 4s^2 + 3s} \right] = 1 = y(0),$$

and the final value is

$$\lim_{s \to 0} \left[s \frac{s^2 + s + 1}{s^3 + 4s^2 + 3s} \right] = 1/3 = y(\infty).$$

You can see that the results from the graphs agree with the results using the initial and the final value theorem, since all the eigenvalues of $sY(s)$ are negative.

The block diagram representing the second system is sketched in Figure 5.21.

EOCE 5.21

Consider the system in Figure 5.22.

If the impulse response is

$$h_1(t) = e^{-t}u(t) = h_2(t),$$

what is the output $y(t)$ if $x(t) = u(t)$?

Find the final and initial values of $y(t)$ and discuss stability.

SOLUTION

$x(t)$ will propagate through the system in Figure 5.22. In the block diagram and between $h_1(t)$ and $h_2(t)$, we can use convolution to write

$$z(t) = x(t) * h_1(t).$$

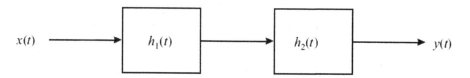

FIGURE 5.22   System for EOCE 5.21.

At the output we write

$$y(t) = z(t) * h_2(t).$$

Taking the Laplace transform of the last two equations we get

$$Z(s) = X(s) * H_1(s)$$

and

$$Y(s) = Z(s) * H_2(s).$$

Substituting $Z(s)$ in $Y(s)$ we get

$$Y(s) = X(s) H_1(s) H_2(s)$$

and

$$H(s) = H_1(s) H_2(s).$$

Therefore, we have

$$Y(s) = \frac{1}{s} \frac{1}{s+1} \frac{1}{s+1} = \frac{1}{s^3 + 2s^2 + s}.$$

We can use MATLAB to solve for $y(t)$ with

$$H(s) = \frac{1}{s^2 + 2s + 1},$$

and by using the script to get

```
poles_of_H =
   -1
   -1
```

The plot is shown in Figure 5.23.

From the plot you can see that the initial value of $y(t)$ is 0 and the final value of $y(t)$ is 1. We also can see that the system is stable (it settles at a final magnitude of 1).

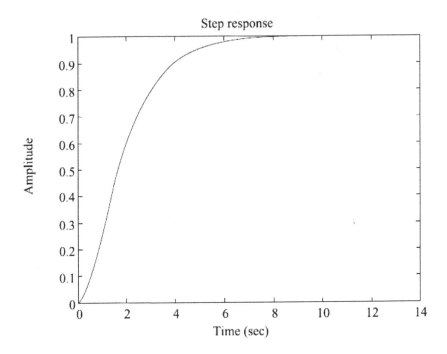

FIGURE 5.23 Plot for EOCE 5.21.

These observations can be confirmed mathematically by looking at the poles of $H(s)$ that are all negative and, therefore, an indication of stability. Also the initial and final values for $y(t)$ can be calculated by using the expression for the output

$$Y(s) = \frac{1}{s} \frac{1}{s^2 + 2s + 1}.$$

The initial value is

$$\lim_{s \to \infty} \left[s \frac{1}{s} \frac{1}{s^2 + 2s + 1} \right] = 0 = y(0),$$

and the final value is

$$\lim_{s \to 0} \left[s \frac{1}{s} \frac{1}{s^2 + 2s + 1} \right] = 1 = y(\infty).$$

EOCE 5.22

Consider the system in Figure 5.24.
    If the impulse response is given by

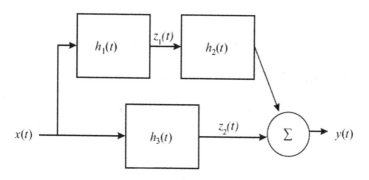

FIGURE 5.24 System for EOCE 5.22.

$$h_1(t) = e^{-t}u(t) = h_2(t) = h_3(t),$$

what is the output $y(t)$ if $x(t) = u(t)$? Find the final and initial values of $y(t)$ and discuss stability.

SOLUTION

$x(t)$ will propagate through the system in Figure 5.24. In the block diagram, and between $h_1(t)$ and $h_2(t)$, we can use convolution to write

$$z_1(t) = x(t) * h_1(t).$$

Taking the Laplace transform we have

$$Z_1(s) = X(s)H_1(s).$$

Before the summer and at the lower part of the block in Figure 5.24 we also have

$$z_2(t) = x(t) * h_3(t)$$

and its transform

$$Z_2(s) = X(s)H_3(s).$$

Before the summer and at the upper part of the block we have

$$z_3(t) = z_1(t) * h_2(t)$$

with its transform

$$Z_3(s) = Z_1(s)H_2(s).$$

At the output we write

$$y(t) = z_2(t) + z_3(t).$$

or

$$Y(s) = Z_2(s) + Z_3(s) = X(s)H_3(s) + X(s)H_1(s)H_2(s);$$

and finally, the transfer function of the system is

$$H(s) = H_3(s) + H_1(s)H_2(s),$$

where

$$H(s) = \frac{1}{s+1} + \frac{1}{s+1}\frac{1}{s+1} = \frac{s+2}{s^2+2s+1}$$

and the output

$$Y(s) = X(s)H(s),$$

or

$$Y(s) = \frac{1}{s}\frac{s+2}{s^2+2s+1}.$$

Now we are ready to use the MATLAB EOCE5_22 script with the output as

```
poles_of_H =
-1
-1
```

The step response is shown in Figure 5.25.

From the plot you can see that the initial value of y(t) is 0 and the final value of y(t) is 2. We also can see that the system is stable (it settles at a final magnitude of 2).

These observations can be confirmed mathematically by looking at the poles of H(s), which are all negative, which is an indication of stability. Also, the initial and final values for y(t) can be calculated with the output written in the Laplace domain as

$$Y(s) = \frac{1}{s}\frac{s+2}{s^3+2s+1}.$$

The initial value of y(t) is

$$\text{limit}_{s \to \infty}\left[s\,\frac{1}{s}\frac{s+2}{s^3+2s+1}\right] = 0 = y(0),$$

and the final value is

$$\text{limit}_{s \to 0}\left[s\,\frac{1}{s}\frac{s+2}{s^3+2s+1}\right] = 2 = y(\infty).$$

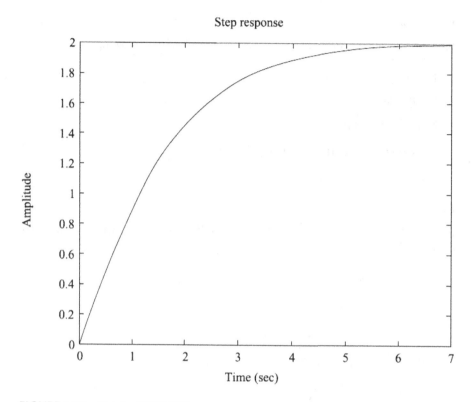

FIGURE 5.25 Plot for EOCE 5.22.

5.14 END-OF-CHAPTER PROBLEMS

EOCP 5.1

Find the Laplace transform for each of the following signals.

1. $2u(t-10)+10\delta(t-1)$
2. $\sin(2(t-1))u(t-1)$
3. $\cos(3(t-5))u(t-5)$
4. $e^{-a(t-b)}u(t-b)+u(t-2)$
5. $(t-1)e^{-a(t-1)}u\,(t-1)$
6. $10e^{-2(t-2)^2}u(t-2)$
7. $2\delta(t-2)+3\delta(t-4)$
8. $e^{-3t}u(t)+\dfrac{t^5}{5!}e^{-10t}u(t)+u(t-5)$

EOCP 5.2

Find the Laplace transform of the derivative for each of the following signals.

1. $tu(t)$
2. $t\sin(t)u(t)$
3. $\left(e^{-3t} + \sin(t)\right)u(t)$
4. $\left(\cos(wt) + \sin(wt)\right)u(t) + 10u(t)$

EOCP 5.3

Find the Laplace transform for each of the following signals, where * indicates convolution.

1. $e^{-t}u(t) * e^{-t}u(t)$
2. $\left[\sin(t)u(t) * u(t)\right] * u(t)$
3. $\left[e^{-t}u(t) + e^{-2t}u(t)\right] * \sin(t)u(t)$
4. $e^{-3t}u(t) * u(t)\right] + u(t) * u(t)$
5. $\delta(t) * \sin(10t)u(t) - \delta(t-1) * u(t)$

EOCP 5.4

Consider the following systems represented by the differential equations. $x(t)$ is the input, and $y(t)$ is the output. Find the total solution.

1. $\dfrac{d}{dt}y + 2y = \sin(t)u(t), \quad y(0) = 1$

2. $\dfrac{d^2}{dt}y + 3\dfrac{d}{dt}y + 2y = u(t), \quad y(0) = 2, \quad \dfrac{d}{dt}y(0) = 0$

3. $\dfrac{d^2}{dt}y + 3\dfrac{d}{dt}y = e^{-t}u(t), \quad y(0) = -1, \quad \dfrac{d}{dt}y(0) = 1$

4. $\dfrac{d^2}{dt}y + 5y = e^{-t}u(t), \quad y(0) = 0, \quad \dfrac{d}{dt}y(0) = -3$

5. $\dfrac{d^2}{dt}y + \dfrac{d}{dt}y + y = \delta(t), \quad y(0) = 2, \quad \dfrac{d}{dt}y(0) = 1$

EOCP 5.5

Given the following Laplace transforms for some signals, what are the time-domain signals?

1. $\dfrac{1}{s+2} + \dfrac{1}{s+3}$

2. $\dfrac{1}{s^2 + 7s + 10}$

3. $\dfrac{1}{s^2 + 5s + 6}$

4. $\dfrac{1}{s^2 + 4s + 4}$

5. $\dfrac{e^{-s}s}{s^2 + 4s + 4}$

6. $\dfrac{e^{-s}s + s}{s^2 + 4s + 4}$

7. $\dfrac{s}{s^2 + 7s + 10}$

8. $\dfrac{s^2}{s^2 + 7s + 10}$

9. $\dfrac{s^3 + 1}{s^2 + 7s + 10}$

10. $\dfrac{s^3 + s^2 + 1}{s^2 + 7s + 10}$

EOCP 5.6

Use the Laplace transform to find $y(t)$, a particular solution, in each of the systems shown in Figures 5.26–5.30.

FIGURE 5.26   System for EOCP 5.6.

FIGURE 5.27   System for EOCP 5.6.

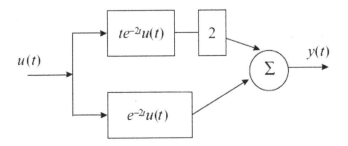

FIGURE 5.28   System for EOCP 5.6.

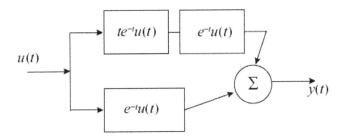

FIGURE 5.29   System for EOCP 5.6.

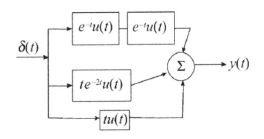

FIGURE 5.30   System for EOCP 5.6.

EOCP 5.7

What is the Laplace transform for each of the signals shown in Figures 5.31–5.34?

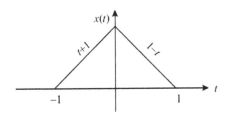

FIGURE 5.31   Signal for EOCP 5.7.

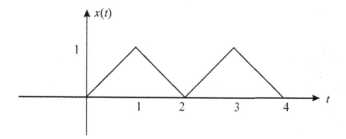

FIGURE 5.32 Signal for EOCP 5.7.

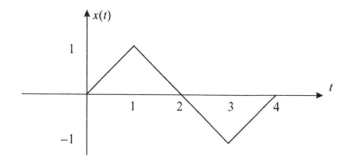

FIGURE 5.33 Signal for EOCP 5.7.

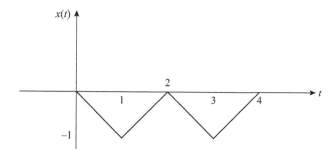

FIGURE 5.34 Signal for EOCP 5.7.

EOCP 5.8

Use the series and parallel models to draw the block diagram for each transfer function.

1. $\dfrac{10}{s^2 - 4}$

2. $\dfrac{10}{2s^2 - 32}$

3. $\dfrac{1}{s^3 + s^2 - 9s - 9}$

4. $\dfrac{s+1}{s^2 + s - 20}$

5. $\dfrac{s^2 + s + 1}{s^3 + 3s^2 - s + 3}$

EOCP 5.9

Draw the block diagram using integrators, summers, and gain blocks for each of the following systems:

1. $\dfrac{s}{s+1}$

2. $\dfrac{2s+1}{s-1}$

3. $\dfrac{2s^2 + s + 1}{s^2 - 1}$

4. $\dfrac{1}{s^2 + 2s + 10}$

5. $\dfrac{s}{s(s+1)}$

6. $\dfrac{100}{2s^2 - 4}$

7. $\dfrac{10}{3s^2 - 16}$

8. $\dfrac{10}{3s^2 + 2s^2 - 9s - 9}$

9. $\dfrac{s^2 + s + 1}{s^2 + s - 20}$

10. $\dfrac{s^2 + s + 1}{5s^2}$

EOCP 5.10

Find the initial and final values for each system using the final and the initial value theorem when applicable.

1. $\dfrac{d^2}{dy} y(t) + 3\dfrac{d}{dt} y(t) + 4y(t) = e^{-t}u(t)$

2. $\dfrac{d^2}{dy} y(t) + 3\dfrac{d}{dt} y(t) - 4y(t) = \sin(t)u(t)$

3. $\dfrac{d^3}{dy}y(t)+3\dfrac{d}{dt}y(t)+4y(t)=e^{-2t}u(t)$

4. $\dfrac{d^3}{dy}y(t)-\dfrac{d}{dt}y(t)=\cos(t)u(t)$

5. $\dfrac{d^3}{dy}y(t)+\dfrac{d^2}{dt}y(t)+\dfrac{d}{dt}y(t)+y(t)=u(t)$

EOCP 5.11

Consider the circuit in Figure 5.35.

1. Find the output voltage $y(t)$, for $t > 0$, using the Laplace transform method. Assume that 10 volts are applied at $t = 0$ seconds.
2. Find the current in the resistor R using the Laplace method.

EOCP 5.12

Consider the mechanical system described by the differential equation and shown in Figure 5.36.

$$M\dfrac{d^2}{dt}y(t)+B\dfrac{d}{dt}y(t)+ky(t)=x(t)$$

The rod to the left of the mass M can be modeled by a translational spring with spring constant k in parallel with a translational damper of constant B.

1. If $x(t)$ is an impulse of strength 10, use the Laplace transform method to find the displacement $y(t)$. Assume no friction and no stored energy.
2. Repeat if $x(t)$ is a sinusoidal shaking of magnitude 10 and period 1 second.

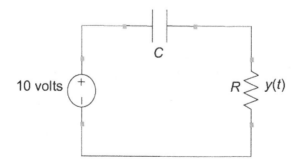

FIGURE 5.35 Circuit for EOCP 5.11.

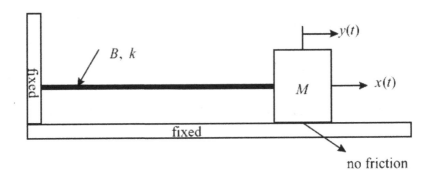

FIGURE 5.36   System for EOCP 5.12.

EOCP 5.13

Consider the circuit in Figure 5.37.

1. If the 10 volts are applied at $t = 0$ seconds, find $y(t)$ using the Laplace transform method.
2. Repeat part 1 but calculate the current in the capacitor.
3. Draw a block diagram of the system using the basic blocks.
4. What are the final and the initial values of the output $y(t)$ and the current in
   the capacitor?

EOCP 5.14

Consider the mechanical system described in Figure 5.38. The masses were initially
displaced. The rod to the left of the mass M_1 can be modeled by a translational spring
with spring constant k_1. The rod to the right of the mass M_1 can be modeled by a
translational spring with spring constant k_2. The rod to the right of the mass M_2 can
be modeled by a translational spring with spring constant k_3. Both masses M_1 and M_2
move smoothly on the fixed surfaces.

FIGURE 5.37   Circuit for EOCP 5.13.

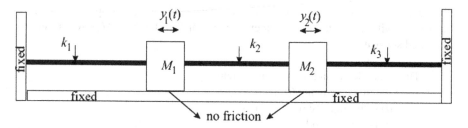

FIGURE 5.38   System for EOCP 5.14.

The two differential equations describing the system are

$$M_1 \frac{d^2}{dt} y_1 + k_1 y_1 + k_2 y_1 - k_2 y_2 = 0$$

$$M_2 \frac{d^2}{dt} y_2 - k_2 y_1 + k_2 y_2 + k_3 y_2 = 0.$$

1. Use the Laplace transform method to find the outputs, $y_1(t)$ and $y_2(t)$.
2. Draw the block diagram for the entire system.

EOCP 5.15

Consider the dynamic system shown in Figure 5.39 and represented by the differential equation

$$M \frac{d^2}{dt} y(t) + (B_1 - B_2) \frac{d}{dt} y(t) + (k_1 - k_2) y(t) = x(t).$$

The rod to the left of the mass M can be modeled by a translational spring with spring constant k_1, in parallel with a translational damper of constant B_1. The rod to

FIGURE 5.39   System for EOCP 5.15.

the right of the mass M can be modeled by a translational spring with spring constant k_2, in parallel with a translational damper of constant B_2. The mass M moves freely on the fixed surface.

1. Draw the block diagram for the system.
2. Find the displacement $y(t)$ if
    a. $x(t)$ is an impulse of unity magnitude
    b. $x(t)$ is a unit step force
3. Find the initial and the final values for $y(t)$.
4. Find the initial and the final values for the velocity of the mass M.

EOCP 5.16

Consider the dynamic system as represented by the differential equation below and shown in Figure 5.40.

$$J\frac{d^2}{dt}\theta(t) + k\theta(t) = T(t)$$

This is a rod attached to a fixed end. Assume that the rod can be represented as a rotational spring with spring constant k, and the mass of the rod is concentrated and modeled by the moment of inertia J.

1. Draw the block diagram representing the system.
2. If the torque is represented as a unit step signal, what is the angular rotation in radians?
3. Find the final and the initial values for $\theta(t)$.

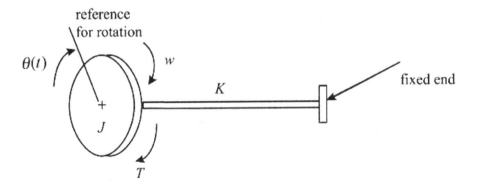

FIGURE 5.40   System for EOCP 5.16.

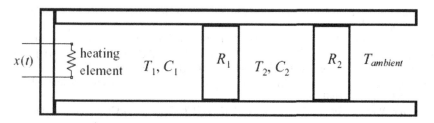

FIGURE 5.41 System for EOCP 5.17.

EOCP 5.17

A dynamic system is given in Figure 5.41.
The differential equations representing the system are

$$\frac{d}{dt}T_1 = \frac{1}{C_1}\left[x(t) - \frac{1}{R_1}(T_1 - T_2)\right]$$

$$\frac{d}{dt}T_2 = \frac{1}{C_2}\left[\frac{1}{R_1}(T_1 - T_2) - \frac{1}{R_2}(T_2 - T)\right],$$

where T_1 and T_2 are the temperatures in the two containers as indicated in Figure 5.41. C_1 and C_2 represent the capacitance in each container. R_1 and R_2 are the resistances of the two separators.

1. Draw the block diagram for the system.
2. If $x(t)$ is a unit step flow of unity magnitude, what are the temperatures $T_1(t)$ and $T_2(t)$?

REFERENCES

Bequette, B.W. *Process Dynamics*, Englewood Cliffs, NJ: Prentice-Hall, 1998.

Brogan, W.L. *Modern Control Theory*, 3rd ed., Englewood Cliffs, NJ: Prentice-Hall, 1991.

Close, M. and Frederick, K. *Modeling and Analysis of Dynamic Systems*, 2nd ed., New York, NY: Wiley, 1995.

Cogdell, J.R. *Foundations of Electrical Engineering*, 2nd ed., Englewood Cliffs, NJ: Prentice-Hall, 1996.

Denbigh, P. *System Analysis and Signal Processing*, Reading, MA: Addison-Wesley, 1998.

Driels, M. *Linear Control System Engineering*, New York, NY: McGraw-Hill, 1996.

Golubitsky, M. and Dellnitz, M. *Linear Algebra and Differential Equations Using MATLAB*, Stamford, CT: Brooks/Cole, 1999.

Harman, T.L., Dabney, J., and Richert, N. *Advanced Engineering Mathematics with MATLAB*, Stamford, CT: Brooks/Cole, 2000.

Kuo, B.C. *Automatic Control System*, 7th ed., Englewood Cliffs, NJ: Prentice-Hall, 1995.

Lewis, P.H. and Yang, C. *Basic Control Systems Engineering*, Englewood Cliffs, NJ: Prentice-Hall, 1997.

Moore, H. *MATLAB for Engineers*, Upper Saddle River, NJ: Prentice-Hall, 2007.

Nilson, W.J. and Riedel, S.A. *Electrical Circuits*, 6th ed., Englewood Cliffs, NJ: Prentice-Hall, 2000.

Nise, N.S. *Control Systems Engineering*, 2nd ed., Reading, MA: Addison-Wesley, 1995.

Ogata, K. *Modern Control Engineering*, 3rd ed., Englewood Cliffs, NJ: Prentice-Hall, 1997.

Ogata, K. *System Dynamics*, 3rd ed., Englewood Cliffs, NJ: Prentice-Hall, 1998.

Phillips, C.L. and Parr, J.M. *Signals, Systems, and Transforms*, 2nd ed., Englewood Cliffs, NJ: Prentice-Hall, 1999.

The MathWorks. *The Student Edition of MATLAB*, Englewood Cliffs, NJ: Prentice-Hall, 1997.

Woods, R.L. and Lawrence, K.L. *Modeling and Simulation of Dynamic Systems*, Englewood Cliffs, NJ: Prentice-Hall, 1997.

Wylie, R.C. and Barrett, C.L. *Advanced Engineering Mathematics*, 6th ed., New York, NY: McGraw-Hill, 1995.

Ziemer, R.E., Tranter, W.H., and Fannin, D.R. *Signals Systems Continuous and Discrete*, 4th ed., Englewood Cliffs, NJ: Prentice-Hall, 1998.

6 State-Space and Linear Systems

6.1 INTRODUCTION

As the interest in many scientific fields increased, modeling systems using linear time-invariant differential equations and tools like transfer functions were not adequate. The state-space approach is superior to other classical methods of modeling. This modern approach can handle systems with nonzero initial conditions (modeling using transfer functions requires that initial conditions be set to zero), as well as time-variant systems. It can also handle linear and nonlinear systems. Up to this point, we have been considering systems with single-input/single-output. The state-space approach can handle multiple-input multiple-output systems.

The state-space approach can also be used to represent or model linear time-invariant single-input/single-output systems. These are the systems we considered in the previous chapters. Systems can have many variables. An example is an electrical circuit where the variables are the inductor current, the capacitor voltage, and the resistor voltage, among others. Using the state-space approach, we will use differential equations to solve for a selected set of these variables. The other variables in the circuit system can be found using the solution for the selected variables. In using the state-space approach, we will follow the procedure detailed below.

We will select specific variables in the system and call them state variables. No state variable selected can be written as a linear combination of the other state variables. Linear combination means that if

$$z_1(t) = 3z_3(t) + 2z_2(t)$$

where $z_1(t)$, $z_2(t)$, and $z_3(t)$ are state variables, we say that $z_1(t)$ is a linear combination of $z_2(t)$ and $z_3(t)$. If we have a first-order differential equation, we will have only one state variable. If the differential equation is second-order, we will have only two state variables. Similarly, if we have an nth-order differential equation, we will have only n state variables. Once we select or decide on the state variables in the system under consideration, we will write a set of first-order simultaneous differential equations, where the right side of these equations is a function only of the state variables (no derivatives) and the inputs to the system, and the number of these equations is determined by the number of state variables selected. We will call this set the state equations set. These state equations will be solved for the selected state variables. All other variables in the system under consideration can be solved using the solutions of these selected state variables and the input to the system. We can use any approach we desire to solve for these selected states. The equations we write to find the other variables in the system are called output equations.

6.2   A REVIEW OF MATRIX ALGEBRA

What follows is a brief review of some of the concepts and definitions we need in this chapter. We will discuss second-order systems when we deal with hand solutions. For matrices of higher dimensions you can consult any linear algebra book.

6.2.1   DEFINITION, GENERAL TERMS, AND NOTATIONS

A matrix is a collection of elements arranged in a rectangular or square array. The size of the matrix is determined by the number of rows and the number of columns in the matrix. A matrix A of m rows and n columns is represented as $A_{m \times n}$. If $m = 1$ then A is a row vector and written as $A_{1 \times n}$. If $n = 1$ then A is a column vector and written as $A_{m \times 1}$. If $n = m$ then A is a square matrix and we write it as $A_{n \times n}$ or $A_{m \times m}$. If all elements in the matrix are zeros we say A is a null matrix or a zero matrix.

6.2.2   THE IDENTITY MATRIX

The identity matrix is the square matrix, where elements along the main diagonal are ones and elements off the main diagonal are zeros. A 2×2 identity matrix is

$$I_{2 \times 2} = \begin{bmatrix} 1 & 0 \\ 0 & 1 \end{bmatrix}. \tag{6.1}$$

6.2.3   ADDING TWO MATRICES

If

$$A = \begin{bmatrix} a & b \\ c & d \end{bmatrix}$$

and

$$B = \begin{bmatrix} e & f \\ g & h \end{bmatrix},$$

then

$$A + B = \begin{bmatrix} a+e & b+f \\ c+g & d+h \end{bmatrix}. \tag{6.2}$$

To add two matrices they must be of the same size. If the matrices are of higher order, the procedure is the same; we add the corresponding entries.

6.2.4 SUBTRACTING TWO MATRICES

If

$$A = \begin{bmatrix} a & b \\ c & d \end{bmatrix}$$

and

$$B = \begin{bmatrix} e & f \\ g & h \end{bmatrix},$$

then

$$A - B = \begin{bmatrix} a-e & b-f \\ c-g & d-h \end{bmatrix}. \tag{6.3}$$

To subtract two matrices they must be of the same size. If the matrices are of higher order, the procedure is the same; we subtract the corresponding entries.

6.2.5 MULTIPLYING A MATRIX BY A CONSTANT

If

$$A = \begin{bmatrix} a & b \\ c & d \end{bmatrix}$$

and k is any given constant, then

$$k \begin{bmatrix} a & b \\ c & d \end{bmatrix} = \begin{bmatrix} ka & kb \\ kc & kd \end{bmatrix}. \tag{6.4}$$

If the matrix A is of higher order, then k is multiplied by each entry in A.

6.2.6 DETERMINANT OF A 2 × 2 MATRIX

Consider the $A_{2\times2}$ matrix

$$A_{2\times2} = \begin{bmatrix} a & b \\ c & d \end{bmatrix}$$

The determinant of A is

$$\det(A) = ad - bc. \tag{6.5}$$

6.2.7 TRANSPOSE OF A MATRIX

If

$$A = \begin{bmatrix} a & b \\ c & d \end{bmatrix},$$

then the transpose of A is given by

$$A^T = \begin{bmatrix} a & c \\ b & d \end{bmatrix}. \tag{6.6}$$

This works for higher order matrices as well, where the first column in A becomes the first row in A^T and so on.

6.2.8 INVERSE OF A MATRIX

If

$$A = \begin{bmatrix} a & b \\ c & d \end{bmatrix},$$

then the inverse of A is

$$A^{-1} = \frac{1}{ad - bc} \begin{bmatrix} d & -b \\ -c & a \end{bmatrix},$$

and since $\dfrac{1}{ad - bc}$ is a constant, we can write the inverse as

$$A^{-1} = \begin{bmatrix} \dfrac{d}{ad - bc} & \dfrac{-b}{ad - bc} \\ \dfrac{-c}{ad - bc} & \dfrac{a}{ad - bc} \end{bmatrix}. \tag{6.7}$$

The inverse of a square matrix exists if the determinant of the matrix is not zero. Also, to find an inverse of a certain matrix, that matrix has to be square. The procedure above for finding the inverse is only for a 2×2 matrix. For higher order matrices, the procedure is different and is found in any linear algebra book.

6.2.9 MATRIX MULTIPLICATION

We can multiply two matrices A and B if the number of columns in A is equal to the number of rows in B.

If $A_{m \times n}$ is to be multiplied by $B_{r \times p}$ then n must be equal to r and the resulting matrix should have m rows and p columns.

If

$$A = \begin{bmatrix} a & b \\ c & d \end{bmatrix}$$

and

$$B = \begin{bmatrix} e & f \\ g & h \end{bmatrix}$$

then if we multiply A by B and let matrix C hold the resulting product. The size of C is 2×2. We could multiply A by B because the number of columns in A, which is two, is equal to the number of rows in B, which is also two. The multiplication of A by B is C and it is

$$C = AB = \begin{bmatrix} a & b \\ c & d \end{bmatrix} \begin{bmatrix} e & f \\ g & h \end{bmatrix}$$

$$C = AB = \begin{bmatrix} ae + bg & af + bh \\ ce + dg & cf + dh \end{bmatrix}. \tag{6.8}$$

We multiply the first row of A, element by element, by all the columns of B. Similarly, we take the second row of A and multiply it by all the columns of B.

Note that in general AB is not the same as BA. The rules for multiplication have to be observed.

6.2.10 Diagonal Form of a Matrix

A matrix A is in diagonal form if all elements in the matrix that are off the diagonal are zeros. If A is not diagonal, we can make it diagonal by finding the matrix P that contains the eigenvectors of A. So if A is not a diagonal matrix, then $(P^{-1}AP)$ will transform A into a diagonal matrix.

6.2.11 Exponent of a Matrix

If A is a square matrix and t is a time variable, then

$$e^{At} = I + At + \frac{(At)^2}{2!} + \frac{(At)^3}{3!} + \frac{(At)^4}{4!} + \cdots \tag{6.9}$$

where I is the identity matrix with the same size as A.

If

$$A = \begin{bmatrix} a & 0 \\ 0 & b \end{bmatrix},$$

a diagonal matrix, then

$$e^{At} = \begin{bmatrix} e^{at} & 0 \\ 0 & e^{bt} \end{bmatrix}.$$

This is true for matrices of bigger sizes, provided that A is a diagonal matrix.

If

$$A = \begin{bmatrix} 0 & 0 \\ 0 & 2 \end{bmatrix},$$

a diagonal matrix also, then

$$e^{At} = \begin{bmatrix} e^{0t} = 1 & 0 \\ 0 & e^{2t} \end{bmatrix}.$$

6.2.12 A SPECIAL MATRIX

Let

$$A = \begin{bmatrix} 0 & a \\ 0 & 0 \end{bmatrix},$$

then

$$A^2 = \begin{bmatrix} 0 & a \\ 0 & 0 \end{bmatrix} \begin{bmatrix} 0 & a \\ 0 & 0 \end{bmatrix} = \begin{bmatrix} 0 & 0 \\ 0 & 0 \end{bmatrix},$$

and

$$A^3 = \begin{bmatrix} 0 & 0 \\ 0 & 0 \end{bmatrix}.$$

A^4, A^5, and so on will have the same value as A^2.

6.2.13   OBSERVATION

Let

$$A = \begin{bmatrix} a & b \\ 0 & a \end{bmatrix}.$$

A can be written as

$$A = \begin{bmatrix} a & 0 \\ 0 & a \end{bmatrix} + \begin{bmatrix} 0 & b \\ 0 & 0 \end{bmatrix}$$

and

$$e^{(At)} = e^{\begin{bmatrix} a & b \\ 0 & a \end{bmatrix} t} = e^{\left\{ \begin{bmatrix} a & 0 \\ 0 & a \end{bmatrix} + \begin{bmatrix} 0 & b \\ 0 & 0 \end{bmatrix} \right\} t}$$

$$e^{(At)} = e^{\begin{bmatrix} a & 0 \\ 0 & a \end{bmatrix} t} e^{\begin{bmatrix} 0 & b \\ 0 & 0 \end{bmatrix} t}.$$

But as we saw earlier, if A is a matrix with all zeros except for the upper right element, then $e^{(At)} = I + (At)$. Knowing this we can finish evaluating the above expression as

$$e^{(At)} = \begin{bmatrix} e^{at} & 0 \\ 0 & e^{at} \end{bmatrix} \left\{ \begin{bmatrix} 1 & 0 \\ 0 & 1 \end{bmatrix} + \begin{bmatrix} 0 & bt \\ 0 & 0 \end{bmatrix} \right\}$$

$$e^{(At)} = \begin{bmatrix} e^{at} & 0 \\ 0 & e^{at} \end{bmatrix} \begin{bmatrix} 1 & bt \\ 0 & 1 \end{bmatrix},$$

and finally,

$$e^{(At)} = \begin{bmatrix} e^{at} & bte^{at} \\ 0 & e^{at} \end{bmatrix}.$$

The above result is very important when we solve state-space first-order equations in real time.

6.2.14   EIGENVALUES OF A MATRIX

The eigenvalues of a matrix A are the roots of the determinant of $(\lambda I - A)$, where I is the identity matrix and λ is a variable.

6.2.15 EIGENVECTORS OF A MATRIX

The eigenvectors of a matrix A are the roots of the homogeneous matrix equation

$$(\lambda I - A)\underline{p} = \underline{0},$$

where $\underline{p}$ is a column vector that represents the eigenvector for a certain eigenvalue.

Example 6.1

Consider the first-order differential equation

$$\frac{d}{dt}y(t) + 2y(t) = x(t),$$

where $y(0^-) = 0$, the initial condition for the output $y(t)$.

Let the input to this system be $x(t) = u(t)$, a unit step input. Find the output $y(t)$ for $t \geq 0$.

SOLUTION

This is a first-order differential equation and therefore we will have one state variable. Let us call that state variable $z_1(t)$.

Next we let

$$z_1(t) = y(t),$$

and then

$$\frac{d}{dt}z_1(t) = \frac{d}{dt}y(t).$$

Rewriting the given differential equation as a function of the state variables we get

$$\frac{d}{dt}z_1 = 2z_1 + x,$$

where again when we write x we mean $x(t)$, the input to the system.

In state-space representation we can write

$$\frac{d}{dt}z_1 = 2z_1 + x$$

as

$$\frac{d}{dt}\underline{z} = A\underline{z} + B\underline{x},$$

where $\underline{z}, \frac{d}{dt}\underline{z}$, and $\underline{x}$ are vectors such that

$$\underline{z} = [z_1], \frac{d}{dt}\underline{z} = \left[\frac{d}{dt}z_1\right] \text{and } \underline{x} = [x].$$

A and B are matrices, where $A = [2]$ and $B = [1]$.

1. The Laplace method
   We can use the Laplace transform to solve this last matrix differential equation by taking the Laplace transform of

$$\frac{d}{dt}\underline{z} = A\underline{z} + B\underline{x}$$

and putting it in a simple algebraic equation in the variable s and write

$$\left[s\underline{Z}(s) - \underline{z}(0^-)\right] = A\underline{Z}(s) + B\underline{X}(s),$$

with $\underline{z}(0^-) = z_1(0^-) = y(0^-) = 0$ and $\underline{X}(s) = 1/s$, since $x(t)$ is a unit step input. Rearranging terms we get

$$\left[s\underline{Z}(s)\right] = A\underline{Z}(s) + B\underline{X}(s) + \underline{z}(0^-),$$

which is a Laplace representation of a first-order single-input single-output system and therefore is one-dimensional. To take $\underline{Z}(s)$ as a common factor in the last equation, $s\underline{Z}(s)$ must be written as $s I \underline{Z}(s)$, where I is the 1×1 identity matrix. Now we write the last equation as

$$[sI - A]\underline{Z}(s) = B\underline{X}(s) + \underline{z}(0^-).$$

By multiplying the above equation by $[sI - A]^{-1}$ to solve for $\underline{Z}(s)$, and substituting for the initial conditions we get

$$\underline{Z}(s) = [sI - A]^{-1} B\underline{X}(s) + [sI - A]^{-1}\underline{z}(0^-)$$

$$\underline{Z}(s) = [s - 2]^{-1}\underline{X}(s) + [s - 2]^{-1}[0].$$

Since $[s - 2]$ is a 1×1 square matrix, its inverse is just its reciprocal. Therefore,

$$\underline{Z}(s) = \left\{1/[s - 2]\right\}\underline{X}(s).$$

With $\underline{X}(s) = U(s)$, the unit step in the Laplace domain, $U(s) = 1/s$. Then

$$\underline{Z}(s) = \left\{1/[s - 2]\right\}\{1/s\}.$$

Using partial fraction expansion we have

$$\underline{Z}(s) = [1]/\left[s(s-2)\right] = [-1/2]/[s] + [1/2]/[s-2].$$

Taking the inverse Laplace transform and solving for $\underline{z}(t)$ we get

$$\underline{z}(t) = [-1/2]u(t) + [1/2]e^{2t}u(t),$$

or

$$\underline{z}(t) = 1/2\left[e^{2t} - 1\right]u(t).$$

$\underline{z}(t)$, which is a 1×1 column vector, is the same as $y(t)$ as defined previously.

2. Real-time method

If we have a differential equation of the form

$$\frac{d}{dt}y(t) + ay(t) = x(t)$$

with initial condition $y(0) = y_0$, then the solution to this equation will have two parts: a solution due to the initial conditions with $x(t) = 0$, the homogeneous part, and a solution due to the input $x(t)$, the particular solution. The solution due to the initial conditions will have the form

$$y_{ic}(t) = y_h(t) = e^{-at}y_0,$$

and the solution due to the input $x(t)$ involves the convolution integral

$$y_p(t) = \int_0^t e^{-a(t-\tau)}x(\tau)\,d\tau.$$

The state-space equation in this example is

$$\frac{d}{dt}\underline{z} = A\underline{z} + B\underline{x},$$

where again, $\underline{z}, \frac{d}{dt}\underline{z}$, and $\underline{x}$ are vectors such that $\underline{z} = [z_1], \frac{d}{dt}\underline{z} = \left[\frac{d}{dt}z_1\right]$
and $\underline{x} = [x]$. A and B are matrices, where $A = [2]$ and $B = [1]$.

The total solution for

$$\frac{d}{dt}\underline{z} = A\underline{z} + B\underline{x}$$

with the initial condition vector $\underline{z}(0^-)$ is

$$\underline{z}(t) = \exp[At]\underline{z}(0) + \int_{t_0}^{t}\exp[A(t-\tau)]Bu(\tau)d\tau,$$

where t_0 in our example is 0 and $u(t)$ is unity in the range from zero to infinity. Notice that A is a 1×1 matrix and therefore is already in a diagonal form. Therefore, the solution with $\underline{z}(0^-) = \underline{0}$ becomes

$$\underline{z}(t) = \int_{0}^{t}\exp[2(t-\tau)]d\tau = \exp(2t)\int_{0}^{t}\exp[-2\tau]d\tau$$

$$\underline{z}(t) = \exp(2t)\frac{\exp[-2\tau]}{-2}\bigg|_{0}^{t} = \frac{1}{-2}\exp(2t)[\exp(-2)-1],$$

and finally,

$$\underline{z}(t) = \frac{1}{2}[\exp(2t)-1]u(t),$$

which is the same solution we get using the Laplace transform technique.

We really do not need to use the state-space approach to solve first-order single-input single-output systems. This example is just to introduce you to the steps needed to deal with problems using the state-space approach.

You will start to feel the power of the state-space approach with the following example.

Example 6.2

Consider the following single-input/single-output linear second-order differential equation

$$\frac{d^2}{dt}y(t) + 3\frac{d}{dt}y(t) + 2y(t) = x(t)$$

with $x(t) = u(t)$ and initial conditions

$$\frac{d}{dt}y(0^-) = y(0^-) = 0.$$

Find the output $y(t)$ for $t \geq 0$.

SOLUTION

In transforming this second-order differential equation into state-space form we need to define two state variables, since the differential equation is given in second order.

Let

$$z_1 = y(t)$$

and

$$z_2 = \frac{d}{dt} y(t).$$

Next, we take derivatives of the two states and write

$$\frac{d}{dt} z_1 = \frac{d}{dt} y(t)$$

and

$$\frac{d}{dt} z_2 = \frac{d^2}{dt} y(t).$$

Substituting for $y(t)$ and $\frac{d}{dt} y(t)$ in the above state equations we get the state equations

$$\frac{d}{dt} z_1 = z_2$$

$$\frac{d}{dt} z_2 = -2z_1 - 3z_2 + x.$$

These are the two simultaneous state equations. Putting them in state-space matrix form with $x(t) = u(t)$, we get

$$\frac{d}{dz} \underline{z} = A\underline{z} + B\underline{x},$$

where

$$A = \begin{bmatrix} 0 & 1 \\ -2 & -3 \end{bmatrix}, B = \begin{bmatrix} 0 \\ 1 \end{bmatrix}, \frac{d}{dt} \underline{z} = \begin{bmatrix} \frac{d}{dt} z_1 \\ \frac{d}{dt} z_2 \end{bmatrix}, \underline{z} = \begin{bmatrix} z_1 \\ z_2 \end{bmatrix}, \text{ and } \underline{x} = [x].$$

The first row in A is obtained by reading the coefficients of z_1 and z_2 in the first state equation. The second row is obtained in a similar way. The first row in B is the coefficient of the input $x(t)$ in the first state equation, and the second row is obtained similarly.

1. The Laplace method

We now try to solve this matrix state equation using the Laplace transform approach. Taking the Laplace transform of the matrix state equation we get

$$\left[s\underline{Z}(s) - \underline{z}(0^-) \right] = A\underline{Z}(s) + B\underline{X}(s).$$

$\underline{z}(0^-) = \underline{0}$, where $\underline{0}$ is a vector of zeros (initial conditions vector).

To take $\underline{Z}(s)$ as a common factor we need to write $s\underline{Z}(s)$ as $sI\ \underline{Z}(s)$, where I is the 2×2 identity matrix to get

$$[sI - A]\underline{Z}(s) = B\underline{X}(s) + \underline{z}(0^-).$$

$[sI - A]$ is a 2×2 matrix. To solve for $\underline{Z}(s)$ we need to multiply the last equation by the inverse of $[sI - A]$ and get

$$\underline{Z}(s) = [sI - A]^{-1} B\underline{X}(s) + [sI - A]^{-1} \underline{z}(0^-).$$

Substituting for A, B, and $\underline{z}(0^-)$ we get

$$\underline{Z}(s) = \left\{ s \begin{bmatrix} 1 & 0 \\ 0 & 1 \end{bmatrix} - \begin{bmatrix} 0 & -1 \\ -2 & -3 \end{bmatrix} \right\}^{-1} \begin{bmatrix} 0 \\ 1 \end{bmatrix} \begin{bmatrix} 1 \\ \frac{1}{s} \end{bmatrix}.$$

This last equation simplifies to

$$\underline{Z}(s) = \begin{bmatrix} s & -1 \\ 2 & s+3 \end{bmatrix}^{-1} \begin{bmatrix} 0 \\ 1/s \end{bmatrix}.$$

Finding the inverse and rewriting we get

$$\underline{Z}(s) = \frac{1}{s^2 + 3s + 2} \begin{bmatrix} s+3 & 1 \\ -2 & s \end{bmatrix} \begin{bmatrix} 0 \\ 1/s \end{bmatrix}$$

or

$$\underline{Z}(s) = \begin{bmatrix} \dfrac{s+3}{s^2 + 3s + 2} & \dfrac{1}{s^2 + 3s + 2} \\ \dfrac{-2}{s^2 + 3s + 2} & \dfrac{s}{s^2 + 3s + 2} \end{bmatrix} \begin{bmatrix} 0 \\ 1/s \end{bmatrix}.$$

Finally,

$$
\underline{Z}(s) = \begin{bmatrix} \dfrac{1}{s\left(s^2 + 3s + 2\right)} \\[4mm] \dfrac{1}{\left(s^2 + 3s + 2\right)} \end{bmatrix}.
$$

Using partial fraction expansion on the transfer function elements in the above matrix, we have

$$
\underline{Z}(s) = \begin{bmatrix} \dfrac{a}{s} + \dfrac{b}{s+1} + \dfrac{c}{s+2} \\[4mm] \dfrac{e}{s+1} + \dfrac{f}{s+2} \end{bmatrix}.
$$

which after calculating the constants gives

$$
\underline{Z}(s) = \begin{bmatrix} \dfrac{1/2}{s} + \dfrac{-1}{s+1} + \dfrac{1/2}{s+2} \\[4mm] \dfrac{1}{s+1} + \dfrac{-1}{s+2} \end{bmatrix}.
$$

Taking the inverse Laplace transform of the above vector, we get

$$
\underline{z}(t) = \begin{bmatrix} Z_1(t) \\[2mm] Z_2(t) \end{bmatrix} = \begin{bmatrix} \dfrac{1}{2}u(t) - e^{-t}u(t) + \dfrac{1}{2}e^{-2t}u(t) \\[3mm] e^{-t}u(t) - e^{-2t}u(t) \end{bmatrix}.
$$

Our goal is to solve for y(t), the output of the system. In this example, y(t) was assigned to the state variable $z_1(t)$. $\dfrac{d}{dt} y(t)$ was assigned to the other state variable $z_2(t)$.

Therefore,

$$
y(t) = \frac{1}{2}u(t) - e^{-t}u(t) + \frac{1}{2}e^{-2t}u(t)
$$

or

$$
y(t) = \left(\frac{1}{2} - e^{-t} + \frac{1}{2}e^{-2t} \right)u(t).
$$

2. Real-time method

The state-space equation is

$$\frac{d}{dt}\underline{z} = A\underline{z} + B\underline{x},$$

where

$$A = \begin{bmatrix} 0 & 1 \\ -2 & -3 \end{bmatrix}, \quad B = \begin{bmatrix} 0 \\ 1 \end{bmatrix}, \quad \frac{d}{dt}\underline{z} = \begin{bmatrix} \frac{d}{dt}z_1 \\ \frac{d}{dt}z_2 \end{bmatrix},$$

$$\underline{z} = \begin{bmatrix} z_1 \\ z_2 \end{bmatrix}, \quad \text{and } \underline{x} = \begin{bmatrix} x \end{bmatrix},$$

with $\underline{z}(t_0) = \underline{z}(0) = 0$, and $x(t) = u(t)$.

The first thing we do here is to notice that A is not in a diagonal form. Hence, we need to diagonalize it. Let us transform the whole system into a new system, where A is transformed into a diagonal matrix $P^{-1}AP$.

Let $\underline{z} = P\,\underline{w}$, where $\underline{w}$ is a new state vector. Take the derivative to get

$$\frac{d}{dt}\underline{z} = P\frac{d}{dt}\underline{w}.$$

By substituting in the state equation

$$\frac{d}{dt}\underline{z} = A\underline{z} + B\underline{x},$$

we get

$$P\frac{d}{dt}\underline{w} = AP\underline{w} + B\underline{x}$$

or

$$\frac{d}{dt}\underline{w} = P^{-1}AP\underline{w} + P^{-1}B\underline{x}.$$

Now this is a new state-space equation. The last equation is obtained by multiplying the equation that precedes it by P^{-1}. The initial condition vector can be transformed into the new space by manipulating the equation

$$\underline{z} = P\underline{w}.$$

Evaluating this equation at $t = 0$ and multiplying it by P^{-1} we arrive at

$$\underline{w}(0) = P^{-1}\underline{z}(0).$$

The total solution now can be written as

$$\underline{w}(t) = \exp\left[P^{-1}APt\right]\underline{w}(0) + \int_0^t \exp\left[P^{-1}AP(t-\tau)\right]P^{-1}Bu(\tau)d\tau.$$

We need to find $P^{-1}AP$ to substitute in the total solution equation.

We start first by finding the eigenvalues of A. We look at the roots of $\det(\lambda I - A)$.

$$\det(\lambda I - A) = \det\left\{\lambda\begin{bmatrix} 1 & 0 \\ 0 & 1 \end{bmatrix} - \begin{bmatrix} 0 & 1 \\ -2 & -3 \end{bmatrix}\right\}$$

$$\det(\lambda I - A) = \det\left\{\lambda\begin{bmatrix} \lambda & -1 \\ 2 & \lambda+3 \end{bmatrix}\right\}$$

which simplifies to

$$\det(\lambda I - A) = \lambda(\lambda+3) - 2(-1)$$

$$\det(\lambda I - A) = \lambda^2 + 3\lambda + 2.$$

The roots of $\det(\lambda I - A)$ are $\lambda_1 = -1$ and $\lambda_2 = -2$.

Now it is time to find the two eigenvectors that correspond to each of the eigenvalues: -1 and -2. To find these vectors we need to find a solution for the equation

$$(\lambda I - A)\underline{p} = \underline{0}$$

for each value of the eigenvalues.

For $\lambda_1 = -1$

$$(\lambda I - A)\underline{p} = \begin{bmatrix} -1 & -1 \\ 2 & 2 \end{bmatrix}\underline{p} = \begin{bmatrix} -1 & -1 \\ 2 & 2 \end{bmatrix}\begin{bmatrix} p_1 \\ p_2 \end{bmatrix} = \underline{0}.$$

The two equations to solve are

$$(-1)p_1 + (-1)p_2 = 0$$

and

$$2p_1 + 2p_2 = 0.$$

These two equations are the same (divide the second by -2 to get the first). To solve them we let $p_1 = c$, then $p_2 = -c$. Therefore, the first eigenvector $\underline{p_1}$ will be

$$\underline{p_1} = c \begin{bmatrix} 1 \\ -1 \end{bmatrix}.$$

Choosing c to have a value of 1 will give

$$\underline{p_1} = \begin{bmatrix} 1 \\ -1 \end{bmatrix},$$

which is the first eigenvector of the matrix P.
   For $\lambda_2 = -2$

$$(\lambda I - A)\underline{p} = \begin{bmatrix} -2 & -1 \\ 2 & 1 \end{bmatrix} \underline{p} = \begin{bmatrix} -2 & -1 \\ 2 & 1 \end{bmatrix} \begin{bmatrix} p_1 \\ p_2 \end{bmatrix} = \underline{0},$$

and the two equations to solve are

$$(-2)p_1 + (-1)p_2 = 0$$

and

$$2p_1 + 1p_2 = 0.$$

These two equations are the same (multiply the second by -1 to get the first). To solve them we let $p_1 = c$, then $p_2 = -2c$. Therefore, the second eigenvector $\underline{p_2}$ will be

$$\underline{p_2} = c \begin{bmatrix} 1 \\ -2 \end{bmatrix}.$$

Choosing c to have a value of 1 will give

$$\underline{p_2} = \begin{bmatrix} 1 \\ -2 \end{bmatrix}.$$

And the matrix P of the two eigenvectors is

$$P = \begin{bmatrix} \underline{p_1} & \underline{p_2} \end{bmatrix} = \begin{bmatrix} 1 & 1 \\ -1 & -2 \end{bmatrix}$$

and

$$P^{-1} = \frac{1}{\det(P)} \begin{bmatrix} -2 & -1 \\ 1 & 1 \end{bmatrix} = -1 \begin{bmatrix} -2 & -1 \\ 1 & 1 \end{bmatrix} = \begin{bmatrix} 2 & 1 \\ -1 & -1 \end{bmatrix}.$$

Now let us look at $P^{-1}AP$ and see if it is really a diagonal matrix.

$$P^{-1}AP = \begin{bmatrix} 2 & 1 \\ -1 & -1 \end{bmatrix} \begin{bmatrix} 0 & 1 \\ -2 & -3 \end{bmatrix} \begin{bmatrix} 1 & 1 \\ -1 & -2 \end{bmatrix}$$

By carrying out the matrix multiplication we arrive at

$$P^{-1}AP = \begin{bmatrix} 2 & -1 \\ 2 & 2 \end{bmatrix} \begin{bmatrix} 1 & 1 \\ -1 & -2 \end{bmatrix} = \begin{bmatrix} -1 & 0 \\ 0 & -2 \end{bmatrix}.$$

Notice that the eigenvalues of A are the same as the eigenvalues of $P^{-1} AP$ (check that out). This means that the shape of the transients is not affected by the transformations we made when we went from the z to the w space.

Now go back to the equation that should give us a solution for $\underline{w}(t)$ and notice that $u(t)$ is the step signal that is unity from time $t = 0$ until infinity.

The solution now can be written as

$$\underline{w}(t) = \exp\left[P^{-1}APt\right]\underline{w}(0) + \int_0^t \exp\left[P^{-1}AP(t-\tau)\right]P^{-1}Bu(\tau)d\tau$$

$$\underline{w}(t) = \int_0^t \begin{bmatrix} e^{-(t-\tau)} & 0 \\ 0 & e^{-2(t-\tau)} \end{bmatrix} \begin{bmatrix} 1 \\ -1 \end{bmatrix} d\tau = \int_0^t \begin{bmatrix} e^{-(t-\tau)} \\ e^{-2(t-\tau)} \end{bmatrix} d\tau.$$

We can integrate element by element to get

$$\underline{w}(t) = \int_0^t \begin{bmatrix} e^{-(t-\tau)} \\ e^{-2(t-\tau)} \end{bmatrix} d\tau = \begin{bmatrix} e^{-t}\left[e^{\tau}\right]_0^t \\ -\dfrac{e^{-2t}}{2}\left[e^{2\tau}\right]_0^t \end{bmatrix}.$$

This simplifies to

$$\underline{w}(t) = \begin{bmatrix} e^{-t}\left[e^{t} - 1\right] \\ \dfrac{1}{-2}e^{-2t}\left[e^{2t} - 1\right] \end{bmatrix} = \begin{bmatrix} 1 - e^{-t} \\ \dfrac{1}{2}\left[e^{-2t} - 1\right] \end{bmatrix}.$$

This is the solution for $\underline{w}(t)$. We need a solution for $\underline{z}(t)$ and therefore we write

$$\underline{z}(t) = P\underline{w}(t)$$

$$\underline{z}(t) = P\underline{w}(t) = \begin{bmatrix} 1 & 1 \\ -1 & -2 \end{bmatrix} \begin{bmatrix} 1 - e^{-t} \\ \dfrac{1}{2}\left[e^{-2t} - 1\right] \end{bmatrix}$$

$$\underline{z}(t) = P\underline{w}(t) = \begin{bmatrix} \dfrac{1}{2} - e^{-t} + \dfrac{1}{2}e^{-2t} \\ \left(e^{-t} - e^{-2t}\right)u(t) \end{bmatrix},$$

from which $y(t)$ is

$$y(t) = \frac{1}{2} - e^{-t} + \frac{1}{2}e^{-2t}.$$

These solutions agree with the solutions obtained using the Laplace transform approach taken earlier in the example.

Example 6.3

Consider the series RLC circuit shown in Figure 6.1, where $x(t)$ is the input.

We are interested in finding the output $v_R(t)$, the voltage across the resistor with $R = 1$ ohm, $C = 1$ farad, and $L = 1$ henry. Let $x(t)$ be an impulse input with unity strength.

SOLUTION

This case is different from the case of a linear constant-coefficient differential equation, where the number of states was set to the order of the differential equation at hand. Now we have a real physical system, the RLC circuit. How do we

FIGURE 6.1   Circuit for Example 6.3.

select the number of states? The system in this example has two energy storage elements (the resistor does not store energy). This means that we will need a second-order differential equation to solve for all variables in the system. Thus, we will have two states. We will select the voltage across the capacitor as one state, and the current in the inductor as the second state, because they relate to the storage elements. We also select the voltage in the capacitor because we know from circuit transient analysis that the voltage in the capacitor does not change instantaneously. We select the current in the inductor because it also does not change instantaneously. Notice that other variables in the circuit were not selected as states. The reason is that the only two variables in the circuit that are linearly independent are the voltage in the capacitor and the current in the inductor. Take, for example, the voltage in the resistor

$$v_R(t) = Ri_L(t),$$

where $i_L(t)$ is the current in the inductor as seen in the circuit. $v_R(t)$ is a linear combination of the current in the inductor. Thus, it is not selected and its value can be calculated easily after we find the current in the inductor. For a circuit that has many resistors, n capacitors that cannot be combined, and m inductors that also cannot be combined, we will have $n + m$ states, where the n states are the n voltages across the capacitors, and the m states are the m currents in the inductors.

Therefore, our two states in this example are the voltage in the capacitor and the current in the inductor, $v_C(t)$ and $i_L(t)$.

In the storage elements,

$$L\frac{d}{dt}i_L(t) = v_L(t)$$

and

$$C\frac{d}{dt}v_C(t) = i_C(t).$$

Remember that the right side of the above state equations must be in terms of the state variables and the input $x(t)$ only.

A loop equation in the circuit in Figure 6.1 will yield the following equations

$$v_L(t) = -Ri_L(t) - v_C(t) + x(t)$$

and

$$i_C(t) = i_L(t),$$

which can be written as

$$L\frac{d}{dt}i_L(t) = -Ri_L(t) - v_C(t) + x(t)$$

and

$$C\frac{d}{dt}v_C(t) = i_L(t).$$

We will let $z_1(t) = i_L(t)$ and $z_2(t) = v_C(t)$. Rewriting the above state equations with $L = 1H$, $C = 1F$, and $R = 1$ ohm we get

$$\frac{d}{dt}z_1(t) = -z_1(t) - z_2(t) + x(t)$$

and

$$\frac{d}{dt}z_2(t) = z_1(t).$$

Assuming zero initial conditions, we can put the two simultaneous state equations in state-space matrix form as

$$\frac{d}{dt}\underline{z} = A\underline{z} + B\underline{x},$$

where

$$A = \begin{bmatrix} -1 & -1 \\ 1 & 0 \end{bmatrix}, \quad B = \begin{bmatrix} 1 \\ 0 \end{bmatrix}, \quad \frac{d}{dt}\underline{z} = \begin{bmatrix} \dfrac{d}{dt}z_1 \\ \dfrac{d}{dt}z_2 \end{bmatrix}, \quad \text{and } \underline{z} = \begin{bmatrix} z_1 \\ z_2 \end{bmatrix}.$$

Following the procedure in the second example, and after taking the Laplace transform of the time-domain state matrix equations, we get

$$\underline{Z}(s) = [sI - A]^{-1}B\underline{X}(s) + [sI - A]^{-1}\underline{z}(0^-).$$

Substituting we get

$$\underline{Z}(s) = \left\{s\begin{bmatrix} 1 & 0 \\ 0 & 1 \end{bmatrix} - \begin{bmatrix} -1 & -1 \\ 1 & 0 \end{bmatrix}\right\}^{-1}\begin{bmatrix} 1 \\ 0 \end{bmatrix}[1],$$

which simplifies to

$$\underline{Z}(s) = \begin{bmatrix} s+1 & 1 \\ -1 & s \end{bmatrix}^{-1} \begin{bmatrix} 1 \\ 0 \end{bmatrix}$$

$$\underline{Z}(s) = \frac{1}{s^2+s+1} \begin{bmatrix} s & -1 \\ 1 & s+1 \end{bmatrix} \begin{bmatrix} 1 \\ 0 \end{bmatrix}$$

$$\underline{Z}(s) = \begin{bmatrix} \dfrac{s}{s^2+s+1} & \dfrac{-1}{s^2+s+1} \\ \dfrac{1}{s^2+s+1} & \dfrac{s+1}{s^2+s+1} \end{bmatrix} \begin{bmatrix} 1 \\ 0 \end{bmatrix}.$$

Finally,

$$\underline{Z}(s) = \begin{bmatrix} \dfrac{s}{s^2+s+1} \\ \dfrac{1}{s^2+s+1} \end{bmatrix}.$$

Taking the inverse Laplace transform of the above state-space Laplace matrix representation we get

$$\underline{z}(t) = \begin{bmatrix} \dfrac{-1}{2}\sqrt{\dfrac{4}{3}}e^{-\frac{1}{2}t}\sin\left(\sqrt{\dfrac{3}{4}}t\right)u(t) + e^{-\frac{1}{2}t}\cos\left(\sqrt{\dfrac{3}{4}}t\right)u(t) \\ \sqrt{\dfrac{4}{3}}e^{-\frac{1}{2}t}\sin\left(\sqrt{\dfrac{3}{4}}t\right)u(t) \end{bmatrix}.$$

The states are

$$z_1(t) = -\frac{1}{2}\sqrt{\frac{4}{3}}e^{-\frac{1}{2}t}\sin\left(\sqrt{\frac{3}{4}}t\right)u(t) + e^{-\frac{1}{2}t}\cos\left(\sqrt{\frac{3}{4}}t\right)u(t)$$

and

$$z_2(t) = \sqrt{\frac{4}{3}}e^{-\frac{1}{2}t}\sin\left(\sqrt{\frac{3}{4}}t\right)u(t).$$

$z_1(t) = i_L(t)$ and $z_2(t) = v_C(t)$. But we were asked to find $v_R(t)$, which is

$$v_R(t) = Ri_L(t).$$

Therefore,

$$v_R(t) = \frac{-1}{2}\sqrt{\frac{4}{3}} e^{-\frac{1}{2}t} \sin\left(\sqrt{\frac{3}{4}}t\right) u(t) + e^{-\frac{1}{2}t} \cos\left(\sqrt{\frac{3}{4}}t\right) u(t).$$

So the input to the system was $x(t) = \delta(t)$, and the output to calculate was $y(t) = v_R(t)$. This example demonstrates how a variable in a system like $v_R(t)$ can be computed from the selected independent states $z_1(t) = i_L(t)$ and $z_2(t) = v_C(t)$. Knowing the states $z_1(t) = i_L(t)$ and $z_2(t) = v_C(t)$, we can find any variable in the circuit. Actually, we can find the following variables:

1. The current in the resistor R: $i_R(t) = i_L(t)$

2. The voltage in the inductor L: $v_L(t) = L\dfrac{d}{dt}i_L(t)$

3. The current in the capacitor C: $i_C(t) = C\dfrac{d}{dt}v_C(t)$

Example 6.4

Consider the following mechanical system with the input $x(t)$ as an impulsive force of unity strength $\delta(t)$ and initial conditions

$$\frac{d}{dy}y(0^-) = y(0^-) = 0.$$

The system is shown in Figure 6.2.

The rod to the left of the mass M can be modeled by a translational spring with spring constant K, in parallel with a translational damper of constant B. $x(t)$ is the input force, and $y(t)$ is the output translation measured from a specific reference as indicated in Figure 6.2. For $M = 1$ kg, $B = 1$ newton/meter/second (N.s/m), $K = 1$ newton/meter. With $x(t) = \delta(t)$, find the output displacement $y(t)$.

FIGURE 6.2 System for Example 6.4.

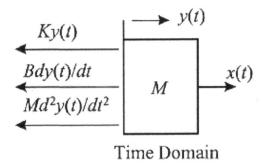

Time Domain

FIGURE 6.3 Free-body diagram for Example 6.4.

SOLUTION

A free-body diagram is constructed to help us write the equations of motion. This diagram is shown in Figure 6.3.

Since we are interested in the output $y(t)$, and if we choose not to use state-space techniques, we can Laplace-transform the free-body diagram in Figure 6.3 and obtain the diagram in Figure 6.4.

Summing the forces using the Laplace-transformed free-body diagram we get

$$KY(s) + BsY(s) + Ms^2Y(s) = X(s).$$

For illustrative purposes only, assuming $M = 1$ kg, $B = 1$ N.s /m, $K = 1$ newton/meter, and with $x(t) = \delta(t)$, we can write

$$Y(s) + sY(s) + s^2Y(s) = 1$$

$$Y(s)\left[s^2 + s + 1\right] = 1.$$

Finally the output is

$$Y(s) = 1/\left[s^2 + s + 1\right] = 1.$$

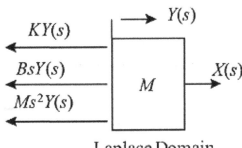

Laplace Domain

FIGURE 6.4 Free-body diagram for Example 6.4.

Taking the inverse Laplace transform of $Y(s)$ we get

$$y(t) = \sqrt{\frac{4}{3}} e^{-\frac{1}{2}t} \sin\left(\sqrt{\frac{3}{4}}t\right) u(t).$$

But we are interested in using the state-space approach in solving for the output $y(t)$. We will consider the free-body diagram in the time domain and sum the forces acting on the object of mass M to get

$$M\frac{d^2}{dt}y(t) + B\frac{d}{dt}y(t) + Ky(t) = x(t).$$

In transforming this second-order differential equation into state-space form, we need to define two state variables, since the differential equation given is second order.

Let

$$z_1 = y(t)$$

and

$$z_2 = \frac{d}{dt}y(t).$$

Next, we take derivatives of the two states to get

$$\frac{d}{dt}z_1 = \frac{d}{dt}y(t)$$

and

$$\frac{d}{dt}z_2 = \frac{d^2}{dt}y(t).$$

In terms of state variables and inputs to the system we have

$$\frac{d}{dt}z_1 = z_2$$

and

$$\frac{d}{dt}z_2 = -z_1 - z_2 + x$$

These are the two simultaneous state equations. Putting them in state-space matrix form with $x(t) = \delta(t)$ we get

$$\frac{d}{dt}\underline{z} = A\underline{z} + B\underline{x},$$

where

$$A = \begin{bmatrix} 0 & 1 \\ -1 & -1 \end{bmatrix}, \quad B = \begin{bmatrix} 0 \\ 1 \end{bmatrix}, \quad \frac{d}{dt}\underline{z} = \begin{bmatrix} \dfrac{d}{dt}z_1 \\ \dfrac{d}{dt}z_2 \end{bmatrix}, \quad \underline{z} = \begin{bmatrix} z_1 \\ z_2 \end{bmatrix}, \quad \text{and } \underline{x} = [x].$$

Up to this point we are still in the time domain. We now try to solve this matrix state equation using the Laplace transform approach. Notice that this system is similar to the system in Example 6.3. We can use the same procedure to get

$$\underline{Z}(s) = \begin{bmatrix} \dfrac{1}{(s^2 + 3s + 2)} \\ \dfrac{s}{(s^2 + 3s + 2)} \end{bmatrix}.$$

Taking the inverse Laplace transform on the above state-space Laplace matrix representation, we get

$$\underline{z}(t) = \begin{bmatrix} \sqrt{\dfrac{4}{3}}\,e^{-\frac{1}{2}t}\sin\left(\sqrt{\dfrac{3}{4}}t\right)u(t) \\ \dfrac{-1}{2}\sqrt{\dfrac{4}{3}}\,e^{-\frac{1}{2}t}\sin\left(\sqrt{\dfrac{3}{4}}t\right)u(t) + e^{-\frac{1}{2}t}\cos\left(\sqrt{\dfrac{3}{4}}t\right)u(t) \end{bmatrix}.$$

Therefore, the states are

$$z_1(t) = \sqrt{\dfrac{4}{3}}\,e^{-\frac{1}{2}t}\sin\left(\sqrt{\dfrac{3}{4}}t\right)u(t)$$

and

$$z_2(t) = \dfrac{-1}{2}\sqrt{\dfrac{4}{3}}\,e^{-\frac{1}{2}t}\sin\left(\sqrt{\dfrac{3}{4}}t\right)u(t) + e^{-\frac{1}{2}t}\cos\left(\sqrt{\dfrac{3}{4}}t\right)u(t).$$

$z_1(t) = y(t)$ and $z_2(t) = \dfrac{d}{dt}y(t)$, which is the velocity of the displacement $y(t)$.

Notice that Examples 6.3 and 6.4 are two different systems in the sense that one is electrical and the other is mechanical, and yet the solutions are similar. The reason is that both systems were modeled mathematically as differential equations.

6.3 GENERAL REPRESENTATION OF SYSTEMS IN STATE SPACE

We will consider systems not only with single-input/single-output, but our study will be general and we will generalize to systems of multiple-inputs/multiple-outputs. We also have indicated that the states we select among the many variables in the system under consideration must be linearly independent. This means that no selected state can be written as a linear combination of the others. A linear combination of n states z_k, where $k = 1$ to n can be represented as

$$\sum_{k=1}^{k=n} \alpha_k z_k = \alpha_1 z_1 + \alpha_2 z_2 + \alpha_3 z_3 + \ldots + \alpha_n z_n, \tag{6.10}$$

where the α's are constants. Further, the selected states are linearly independent if the sum

$$\sum_{k=1}^{k=n} \alpha_k z_k = \alpha_1 z_1 + \alpha_2 z_2 + \alpha_3 z_3 + \ldots + \alpha_n z_n$$

is equal to zero only if all the α's are zero and no selected state is zero.

System variables are the variables in the system where, when an input is applied to the system, they will have a response whether this input is an actual external input or an internal input (initial condition).

The state variables of the system are the minimum linearly independent set of variables in the same system.

When we express the output variables in the system as a linear combination of the state variables, these output variables are called output equations for the system. In a similar way, we say that the set of the first-order differential equations that describe the system completely are called state equations. When we group our state variables in a vector form, we will call them state vectors. The n-dimensional space whose variables are the state variables is called the state-space. Multiple-input/multiple-output systems can be represented in state space as the two state-space matrix equations

$$\frac{d}{dt}\underline{z} = A\underline{z} + B\underline{x} \tag{6.11}$$

$$\underline{y} = C\underline{z} + D\underline{x}.$$

In the above equation, $\underline{z}$ is the state vector, A is the system coefficient matrix, B is the input matrix, C is the output matrix, and D is the feed-forward matrix.

The first of the equations above is called the state equation, and the second is called the output equation.

6.4   GENERAL SOLUTION OF STATE-SPACE EQUATIONS USING THE LAPLACE TRANSFORM

Given the state-space matrix equations in the time domain as

$$\frac{d}{dt}\underline{z} = A\underline{z} + B\underline{x}$$

$$\underline{y} = C\underline{z} + D\underline{x},$$

the solution for these equations in the Laplace domain is

$$\underline{Z}(s) = (sI - A)^{-1}\underline{z}(0) + (sI - A)^{-1} B\underline{X}(s) \qquad (6.12)$$

$$\underline{Y}(s) = C\underline{Z}(s) + D\underline{X}(s).$$

6.5   GENERAL SOLUTION OF THE STATE-SPACE EQUATIONS IN REAL TIME

Given the state-space matrix equations in the time domain as

$$\frac{d}{dt}\underline{z} = A\underline{z} + B\underline{x}$$

$$\underline{y} = C\underline{z} + D\underline{x},$$

we will examine the two cases, where the matrix A can have two different forms.

If A is a diagonal matrix, then $\exp(At)$ is straightforward and explained in the examples above. The solution for the states in this case is

$$\underline{z}(t) = \exp[At]\underline{z}(0) + \int_0^t \exp[A(t-\tau)]B\underline{x}(\tau)d\tau, \qquad (6.13)$$

and for the outputs it is

$$\underline{y}(t) = C\underline{z}(t) + D\underline{x}(t). \qquad (6.14)$$

If A is not a diagonal matrix, it can be diagonalized using the techniques explained in the previous examples, and the representation in state space can be transformed to a new space $\underline{w}(t)$ as a result. The solution for the states is then

$$\underline{z}(t) = P\underline{w}(t), \qquad (6.15)$$

and for the outputs it is

$$\underline{y}(t) = CP\underline{w}(t) + D\underline{x}(t), \qquad (6.16)$$

where

$$\underline{w}(t) = \exp\left[P^{-1}APt\right]\underline{w}(0) + \int_0^t \exp\left[P^{-1}AP(t-\tau)P^{-1}Bx(\tau)d\tau.\right] \quad (6.17)$$

P is the transformation matrix that transforms A into $P^{-1}AP$, the diagonalized matrix.

6.6 WAYS OF EVALUATING e^{At}

There are many ways of evaluating the exponent of the matrix A. We will consider six of them in this chapter.

6.6.1 FIRST METHOD: A IS A DIAGONAL MATRIX

If A is diagonal square matrix, then e^{At} is a square matrix, too. It is calculated by leaving alone all elements in A, which are zeros and using the other elements multiplied by t as exponents. As an example:
   If

$$A = \begin{bmatrix} a & 0 \\ 0 & b \end{bmatrix}$$

is a diagonal matrix, then

$$e^{At} = \begin{bmatrix} e^{at} & 0 \\ 0 & e^{bt} \end{bmatrix}. \quad (6.18)$$

6.6.2 SECOND METHOD: A IS OF THE FORM $\begin{bmatrix} a & b \\ 0 & a \end{bmatrix}$

This was shown previously and is presented here for completeness as

$$\exp(At) = \begin{bmatrix} e^{at} & bte^{at} \\ 0 & e^{at} \end{bmatrix}. \quad (6.19)$$

6.6.3 THIRD METHOD: NUMERICAL EVALUATION, A OF ANY FORM

From Taylor's series, we can write the exponent of the matrix A as

$$e^{At} = I + At + \frac{(At)^2}{2!} + \frac{(At)^3}{3!} + \frac{(At)^4}{4!} + \dots \quad (6.20)$$

This method is not desirable if a complete closed form solution is sought.

6.6.4 FOURTH METHOD: THE CAYLEY–HAMILTON APPROACH

Using the Cayley–Hamilton theorem, we can write

$$e^{At} = a_0 I + a_1 A + a_2 A^2 + a_3 A^3 + \cdots + a_{n-1} A^{n-1}, \tag{6.21}$$

where a_0 through a_{n-1} are to be calculated, and n is the size of the matrix A. Let us denote the eigenvalues for A as λ_1 through λ_n, and substitute A by each of the eigenvalues to get the following set of equations

$$e^{\lambda_1 t} = a_0 + a_1 \lambda_1 + a_2 \lambda_1^2 + \cdots + a_{n-1} \lambda_1^{n-1}$$

$$e^{\lambda_2 t} = a_0 + a_1 \lambda_2 + a_2 \lambda_2^2 + \cdots + a_{n-1} \lambda_2^{n-1}$$

$$e^{\lambda_3 t} = a_0 + a_1 \lambda_3 + a_2 \lambda_3^2 + \cdots + a_{n-1} \lambda_3^{n-1}$$

$$\vdots$$

$$e^{\lambda_n t} = a_0 + a_1 \lambda_n + a_2 \lambda_n^2 + \cdots + a_{n-1} \lambda_n^{n-1} \tag{6.22}$$

If the eigenvalues of A are distinct, we will have n simultaneous equations to solve for the coefficients a_0 through a_{n-1}. If the eigenvalues of A are not distinct, then the above set of equations will be dependent. If an eigenvalue has, for example, a multiplicity of two, then we will need an extra equation involving the derivative of the equation corresponding to that particular eigenvalue with respect to the eigenvalue itself. As an example, consider

$$A = \begin{bmatrix} -2 & 0 \\ 0 & -2 \end{bmatrix}.$$

The characteristic equation $(\det(\lambda I - A) = 0)$ is

$$\lambda^2 + 4\lambda + 4 = 0.$$

The roots are at $\lambda = -2$ and $\lambda = -2$.

Using the general equations presented before, we get

$$e^{\lambda t} = a_0 + \lambda a_1$$

and its derivative

$$t e^{\lambda t} = a_1.$$

With the eigenvalues both at -2, the above equations become

$$e^{-2t} = a_0 - 2a_1$$

$$te^{-2t} = a_1.$$

Therefore,

$$a_1 = te^{-2t}$$

$$a_0 = 2te^{-2t} + e^{-2t}.$$

Since A is a 2×2 matrix, we have

$$e^{At} = a_0 I + a_1 A$$

$$e^{At} = a_0 I + a_1 A = \begin{bmatrix} a_0 - 2a_1 & 0 \\ 0 & a_0 - 2a_1 \end{bmatrix} = \begin{bmatrix} e^{-2t} & 0 \\ 0 & e^{-2t} \end{bmatrix}.$$

As another example, let

$$A = \begin{bmatrix} 0 & 1 \\ -6 & -5 \end{bmatrix}.$$

We will first find the eigenvalues for A. The characteristic equation is

$$\lambda^2 + 5\lambda + 6 = 0.$$

The roots are at $\lambda = -2$ and $\lambda = -3$.

Using the general equations presented before, we write

$$e^{-2t} = a_0 - 2a_1$$

$$e^{-3t} = a_0 - 3a_1.$$

Solving these simultaneous equations leads to

$$a_0 = -2e^{-3t} + 3e^{-2t}$$

$$a_1 = e^{-2t} - e^{-3t}.$$

Since A is a 2×2 matrix we have

$$e^{At} = a_0 I + a_1 A$$

$$e^{At} = a_0 I + a_1 A = \begin{bmatrix} a_0 & a_1 \\ -6a_1 & a_0 - 5a_1 \end{bmatrix} = \begin{bmatrix} 3e^{-2t} - 2e^{-2t} & e^{-2t} - e^{-3t} \\ -6e^{-2t} + 6e^{-3t} & -2e^{-2t} + 3e^{-3t} \end{bmatrix}.$$

6.6.5 FIFTH METHOD: THE INVERSE LAPLACE METHOD

e^{At} is known as the state-transition matrix and usually denoted as $\Phi(t)$. Let us look at the Laplace transform solution of the state equations

$$\underline{Z}(s) = (sI - A)^{-1} \underline{z}(0) + (sI - A)^{-1} B\underline{X}(s),$$

and for comparison, let us give the time-domain solution for the same equations

$$\underline{z}(t) = \exp[At]\underline{z}(0) + \int_0^t \exp[A(t-\tau)]Bx(\tau)d\tau.$$

A close look at these equations tells us that e^{At} is the inverse Laplace transform of $(sI - A)^{-1}$. Or mathematically, we write

$$e^{At} = L^{-1}\left[(sI - A)^{-1}\right]. \tag{6.23}$$

As an example, let

$$A = \begin{bmatrix} -2 & 0 \\ 0 & -2 \end{bmatrix}.$$

Then

$$(sI - A)^{-1} = \begin{bmatrix} s+2 & 0 \\ 0 & s+2 \end{bmatrix}^{-1} = \frac{1}{(s+2)^2} \begin{bmatrix} s+2 & 0 \\ 0 & s+2 \end{bmatrix}$$

$$(sI - A)^{-1} = \begin{bmatrix} \dfrac{1}{s+2} & 0 \\ 0 & \dfrac{1}{s+2} \end{bmatrix}.$$

Taking the inverse transform we get

$$
\begin{bmatrix} e^{-2t} & 0 \\ 0 & e^{-2t} \end{bmatrix} = e^{At} = \Phi(t).
$$

Before we go into the sixth method of finding e^{At}, we will look at some important properties of the state-transition matrix.

We know that for systems with zero input

$$
\underline{z}(t) = \Phi(t)\underline{z}(0).
$$

Let $t = 0$. Then

$$
\underline{z}(0) = \Phi(0)\underline{z}(0).
$$

says that

$$
\Phi(0) = I,
$$

which is the identity matrix. Let us differentiate

$$
\underline{z}(t) = \Phi(t)\underline{z}(0)
$$

with respect to time. We will get

$$
\frac{d}{dt}\underline{z}(t) = \frac{d}{dt}\Phi(t)\underline{z}(0) = A\underline{z}(t).
$$

Let us set $t = 0$ in the last equation to get

$$
\frac{d}{dt}\Phi(0)\underline{z}(0) = A\underline{z}(0).
$$

This last equation tells us that

$$
\frac{d}{dt}\Phi(0) = A.
$$

6.6.6 SIXTH METHOD: USING THE GENERAL FORM OF $\Phi(t) = e^{At}$ AND ITS PROPERTIES

We also noted that each term in the transition matrix is generated by the system poles. So, in general, for a 2×2 system with poles λ_1 and λ_2, the form of the transition matrix is

$$
\Phi(t) = \begin{bmatrix} a_1 e^{\lambda_1 t} + a_2 e^{\lambda_2 t} & a_3 e^{\lambda_1 t} + a_4 e^{\lambda_2 t} \\ a_5 e^{\lambda_1 t} + a_6 e^{\lambda_2 t} & a_7 e^{\lambda_1 t} + a_8 e^{\lambda_2 t} \end{bmatrix},
$$

where the constants a_1 through a_8 are to be determined.

Let us consider an example to illustrate the procedure.
Let

$$A = \begin{bmatrix} 0 & 1 \\ -8 & -6 \end{bmatrix}.$$

Since A is of dimension 2×2,

$$\Phi(t) = \begin{bmatrix} a_1 e^{\lambda_1 t} + a_2 e^{\lambda_2 t} & a_3 e^{\lambda_1 t} + a_4 e^{\lambda_2 t} \\ a_5 e^{\lambda_1 t} + a_6 e^{\lambda_2 t} & a_7 e^{\lambda_1 t} + a_8 e^{\lambda_2 t} \end{bmatrix}. \tag{6.24}$$

We will use the properties of the transition matrix to find the constants a_1 to a_8.

$$\Phi(0) = \begin{bmatrix} a_1 + a_2 & a_3 + a_4 \\ a_5 + a_6 & a_7 + a_8 \end{bmatrix} = \begin{bmatrix} 1 & 0 \\ 0 & 1 \end{bmatrix} \tag{6.25}$$

$$\frac{d}{dt}\Phi(t) = \begin{bmatrix} \lambda_1 a_1 e^{\lambda_1 t} + \lambda_2 a_2 e^{\lambda_2 t} & \lambda_1 a_3 e^{\lambda_1 t} + \lambda_2 a_4 e^{\lambda_2 t} \\ \lambda_1 a_5 e^{\lambda_1 t} + \lambda_2 a_6 e^{\lambda_2 t} & \lambda_1 a_7 e^{\lambda_1 t} + \lambda_2 a_8 e^{\lambda_2 t} \end{bmatrix} \tag{6.26}$$

Evaluating at $t = 0$, we get

$$\frac{d}{dt}\Phi(0) = \begin{bmatrix} \lambda_1 a_1 + \lambda_2 a_2 & \lambda_1 a_3 + \lambda_2 a_4 \\ \lambda_1 a_5 + \lambda_2 a_6 & \lambda_1 a_7 + \lambda_2 a_8 \end{bmatrix} = \begin{bmatrix} 0 & 1 \\ -8 & -6 \end{bmatrix}.$$

We will have eight algebraic equations to solve. These equations are

$$a_1 + a_2 = 1$$
$$a_3 + a_4 = 0$$
$$a_5 + a_6 = 0 \tag{6.27}$$
$$a_7 + a_8 = 1$$

and

$$\lambda_1 a_1 + \lambda_2 a_2 = 0$$
$$\lambda_1 a_3 + \lambda_2 a_4 = 1$$
$$\lambda_1 a_5 + \lambda_2 a_6 = -8 \tag{6.28}$$
$$\lambda_1 a_7 + \lambda_2 a_8 = -6.$$

With λ_1 and λ_2 both at -2 and -4, we can solve Equation 1 of set (Equation 6.27) and Equation 1 of set (Equation 6.28) and so forth to get all the constant terms. The constants are $a_1 = 2$, $a_2 = -1$, $a_3 = 0.5$, $a_4 = -0.5$, $a_5 = -4$, $a_6 = 4$, $a_7 = -1$, and $a_8 = 2$. Finally, the transition matrix $\Phi(t)$ is

$$\Phi(t) = \begin{bmatrix} 2e^{\lambda_1 t} - e^{\lambda_2 t} & 0.5e^{\lambda_1 t} - 0.5e^{\lambda_2 t} \\ -4e^{\lambda_1 t} + 4e^{\lambda_2 t} & -e^{\lambda_1 t} + 2e^{\lambda_2 t} \end{bmatrix}.$$

Example 6.5

Consider the state-space system represented as

$$\frac{d}{dt}\underline{z} = \begin{bmatrix} 0 & 1 \\ -6 & -5 \end{bmatrix}\underline{z}$$

for the states, and

$$\underline{y} = \begin{bmatrix} 0 & 1 \\ 1 & 0 \end{bmatrix}\underline{z}$$

for the outputs.
With the initial conditions

$$\underline{z}(0) = \begin{bmatrix} 0 \\ 1 \end{bmatrix},$$

what are the outputs, y_1 and y_2? Calculate the state-transition matrix using the methods presented neglecting the numerical method.

SOLUTION

1. Using the Cayley–Hamilton method
   Note that this A matrix is the same matrix we used when we presented the Cayley–Hamilton method. We will present the transition matrix here for completeness.

$$e^{At} = a_0 I + a_1 A$$

which is finally

$$e^{At} = \begin{bmatrix} 3e^{-2t} - 2e^{-3t} & e^{-2t} - e^{-3t} \\ -6e^{-2t} + 6e^{-3t} & -2e^{-2t} + 3e^{-3t} \end{bmatrix}.$$

2. Using the inverse transform method

First we find

$$(sI - A)^{-1} = \begin{bmatrix} s & -1 \\ 6 & s+5 \end{bmatrix}^{-1} = \frac{1}{(s^2 + 5s + 6)} \begin{bmatrix} s+5 & 1 \\ -6 & s \end{bmatrix},$$

which simplifies to

$$(sI - A)^{-1} = \begin{bmatrix} \dfrac{s+5}{(s+2)(s+3)} & \dfrac{1}{(s+2)(s+3)} \\ \dfrac{-6}{(s+2)(s+3)} & \dfrac{s}{(s+2)(s+3)} \end{bmatrix}.$$

Using partial fraction expansion, we can write

$$(sI - A)^{-1} = \begin{bmatrix} \dfrac{3}{(s+2)} - \dfrac{2}{(s+3)} & \dfrac{1}{(s+2)} - \dfrac{1}{(s+3)} \\ \dfrac{-6}{(s+2)} + \dfrac{6}{(s+3)} & \dfrac{-2}{(s+2)} + \dfrac{3}{(s+3)} \end{bmatrix}.$$

Taking the inverse transform using partial fraction expansion, we will get $\Phi(t)$ as

$$\Phi(t) = \begin{bmatrix} 3e^{-2t} - 2e^{-3t} & e^{-2t} - e^{-3t} \\ -6e^{-2t} + 6e^{-3t} & -2e^{-2t} + 3e^{-3t} \end{bmatrix}.$$

3. Using the general form of the transition matrix and its properties. Since A is of dimension 2×2,

$$\Phi(t) = \begin{bmatrix} a_1 e^{\lambda_1 t} + a_2 e^{\lambda_2 t} & a_3 e^{\lambda_1 t} + a_4 e^{\lambda_2 t} \\ a_5 e^{\lambda_1 t} + a_6 e^{\lambda_2 t} & a_7 e^{\lambda_1 t} + a_8 e^{\lambda_2 t} \end{bmatrix}.$$

By evaluating at $t = 0$ we get

$$\Phi(0) = \begin{bmatrix} a_1 + a_2 & a_3 + a_4 \\ a_5 + a_6 & a_7 + a_8 \end{bmatrix} = \begin{bmatrix} 1 & 0 \\ 0 & 1 \end{bmatrix}$$

and

$$\frac{d}{dt}\Phi(0) \begin{bmatrix} \lambda_1 a_1 + \lambda_2 a_2 & \lambda_1 a_3 + \lambda_2 a_4 \\ \lambda_1 a_5 + \lambda_2 a_6 & \lambda_1 a_7 + \lambda_2 a_8 \end{bmatrix} = \begin{bmatrix} 0 & 1 \\ -6 & -5 \end{bmatrix}.$$

Again, we will have eight algebraic equations to solve. These equations are

$$a_1 + a_2 = 1$$

$$a_3 + a_4 = 0$$

$$a_5 + a_6 = 0 \tag{6.29}$$

$$a_7 + a_8 = 1$$

and

$$\lambda_1 a_1 + \lambda_2 a_2 = 0$$

$$\lambda_1 a_3 + \lambda_2 a_4 = 1$$

$$\lambda_1 a_5 + \lambda_2 a_6 = -6 \tag{6.30}$$

$$\lambda_1 a_7 + \lambda_2 a_8 = -5.$$

With λ_1 and λ_2 both at -2 and -3, we can solve Equation 1 of set (Equation 6.29) and Equation 1 of set (Equation 6.30) and so forth to get all the constant terms. In doing that we find $a_1 = 3$, $a_2 = -2$, $a_3 = 1$, $a_4 = -1$, $a_5 = -6$, $a_6 = 6$, $a_7 = -2$, and $a_8 = 3$. With these values the transition matrix $\Phi(t)$ is obtained as

$$\Phi(t) = \begin{bmatrix} a_1 e^{\lambda_1 t} + a_2 e^{\lambda_2 t} & a_3 e^{\lambda_1 t} + a_4 e^{\lambda_2 t} \\ a_5 e^{\lambda_1 t} + a_6 e^{\lambda_2 t} & a_7 e^{\lambda_1 t} + a_8 e^{\lambda_2 t} \end{bmatrix}$$

$$= \begin{bmatrix} 3e^{\lambda_1 t} - 2e^{\lambda_2 t} & e^{\lambda_1 t} - e^{\lambda_2 t} \\ -6e^{\lambda_1 t} + 6e^{\lambda_2 t} & -2e^{\lambda_1 t} + 3e^{\lambda_2 t} \end{bmatrix}$$

The solution for the states is

$$\underline{z} = \begin{bmatrix} 3e^{-2t} - 2e^{-3t} & e^{-2t} - e^{-3t} \\ -6e^{-2t} + 6e^{-3t} & -2e^{-2t} + 3e^{-3t} \end{bmatrix} \begin{bmatrix} 0 \\ 1 \end{bmatrix}$$

$$= \begin{bmatrix} e^{-2t} - e^{-3t} \\ -2e^{-2t} + 3e^{-3t} \end{bmatrix},$$

and for the outputs

$$\underline{y} = \begin{bmatrix} 0 & 1 \\ 1 & 0 \end{bmatrix} \begin{bmatrix} e^{-2t} - e^{-3t} \\ -2e^{-2t} + 3e^{-3t} \end{bmatrix} = \begin{bmatrix} -2e^{-2t} + 3e^{-3t} \\ e^{-2t} - e^{-3t} \end{bmatrix}.$$

For hand examples, the inverse Laplace method is the fastest for second-order systems.

6.7   SOME INSIGHTS: POLES AND STABILITY

The objective of this chapter was to represent linear systems in state-space form and to look for ways of solving for the states. The process was to represent an nth-order system (nth-order differential equation) as n first-order differential equations and arrange these equations in what we call state-space representation as

$$\frac{d}{dt}\underline{z} = A\underline{z} + B\underline{x}$$

$$\underline{y} = C\underline{z} + D\underline{x},$$

where $\underline{x}$ is the input vector (assuming multiple inputs), and $\underline{y}$ is the output vector (assuming multiple outputs). The $\underline{z}$ vector is the vector that contains the states of the system. The A matrix is the matrix that contains the parameters that control the dynamics of the system. As we saw in previous chapters, in every system representation there was a way to find the eigenvalues of the system. In state-space representation, the roots of the determinant of the matrix, $(sI - A)$, where I is the identity matrix, are the eigenvalues of the system, the poles. And as we mentioned before, these poles determine the shape of the transients of the system under investigation.

Consider the following single-input/single-output linear second-order differential equation that we discussed in Example 6.2 earlier in this chapter

$$\frac{d^2}{dt}y(t) + 3\frac{d}{dt}y(t) + 2y(t) = x(t)$$

with $x(t) = u(t)$ and initial conditions

$$\frac{d}{dt}y(0^-) = y(0^-) = 0.$$

Putting this differential equation in state-space matrix form with $x(t) = u(t)$, we get

$$\frac{d}{dt}\underline{z} = A\underline{z} + B\underline{x},$$

where

$$A = \begin{bmatrix} 0 & 1 \\ -2 & -3 \end{bmatrix}, \ B = \begin{bmatrix} 0 \\ 1 \end{bmatrix}, \ \frac{d}{dt}\underline{z} = \begin{bmatrix} \dfrac{d}{dt}z_1 \\ \dfrac{d}{dt}z_2 \end{bmatrix}, \ \underline{z} = \begin{bmatrix} z_1 \\ z_1 \end{bmatrix}, \text{ and } \underline{x} = [x].$$

The roots of the determinant of $(sI - A)$ are calculated as

$$sI - A = s \begin{bmatrix} 1 & 0 \\ 0 & 1 \end{bmatrix} - \begin{bmatrix} 0 & 1 \\ -2 & -3 \end{bmatrix} = \begin{bmatrix} s & -1 \\ 2 & s+3 \end{bmatrix}.$$

The determinant of the above matrix is

$$s^2 + 3s + 2 = (s+1)(s+2).$$

The roots of the above equation are at $s = -1$ and at $s = -2$. These are the eigenvalues or the poles of the system. Therefore, as we mentioned earlier, we expect a solution that contains the terms $c_1 e^{-1t} + c_2 e^{-2t}$. As you can see, this solution is stable (eigenvalues are all negative). Let us look at the solution we arrived at earlier

$$y(t) = \left(1 + \frac{1}{2}e^{-t} - e^{-2t}\right)u(t).$$

You can see that $c_1 = 1/2$ and $c_2 = -1$. These coefficients have no effect on the stability of the system. The extra term in the output $y(t)$ is the term $u(t)$. This term is due to the input $x(t)$. So if the input is bounded, and the eigenvalues are all negative, the output has to be stable; it dies as time progresses or settles at a certain bounded value, in this case $u(t)$.

To summarize, if the system is given in state-space form, the stability of the system is determined by finding the roots of $\det[sI - A]$. If the roots are all negative (in case of complex roots we look at the real part of the roots), the system is stable. If one of the roots is positive, the system is unstable. And again, the roots will determine the shape of the transients.

6.8 END-OF-CHAPTER EXAMPLES

EOCE 6.1

Consider the following two matrices:

$$A = \begin{bmatrix} -14 & 4 \\ 1 & -18 \end{bmatrix}$$

and

$$B = \begin{bmatrix} 2 & 1 \\ 3 & 0 \end{bmatrix}.$$

We will utilize MATLAB to find each of the following:
$A + B$, $B - A$, $23A$, $2(A + B)$, AB, BA, $2AB$, transpose(A), and inverse(A).

SOLUTION

Let us write the script EOCE6_1.

```
The output is then
sum_A_and_B=
```

$$
\begin{array}{rr}
-12 & 5 \\
4 & -18
\end{array}
$$

```
Diff_B_A=
```

$$
\begin{array}{rr}
16 & -3 \\
2 & 18
\end{array}
$$

```
A_times_23=
```

$$
\begin{array}{rr}
-322 & 92 \\
23 & -414
\end{array}
$$

```
A_plus_B_times_2=
```

$$
\begin{array}{rr}
-24 & 10 \\
8 & -36
\end{array}
$$

```
A_times_B=
```

$$
\begin{array}{rr}
-16 & -14 \\
-52 & 1
\end{array}
$$

```
B_times_A=
```

$$
\begin{array}{rr}
-27 & -10 \\
-42 & 12
\end{array}
$$

```
Two_times_A_times_B=
```

$$
\begin{array}{rr}
-32 & -28 \\
-104 & 2
\end{array}
$$

```
Transpose_A=
```

$$\begin{array}{cc} -14 & 1 \\ 4 & -18 \end{array}$$

```
Inverse_A=
```

$$\begin{array}{cc} -0.0726 & -0.0161 \\ -0.0040 & -0.0565 \end{array}$$

In the above example, notice that AB is not equal to BA. This is to say that multiplication may not commute.

EOCE 6.2

Consider the two matrices given in EOCE 6.1 again.

$$A = \begin{bmatrix} -14 & 4 \\ 1 & -18 \end{bmatrix} \text{ and } B = \begin{bmatrix} 2 & 1 \\ 3 & 0 \end{bmatrix}.$$

Find the eigenvalues and eigenvectors of the above two matrices using MATLAB.

SOLUTION

To do that we write the script
EOCE6_2. The output is

```
VA=
```

$$\begin{array}{cc} 0.9792 & -0.6380 \\ 0.2028 & 0.7701 \end{array}$$

```
DA=
```

$$\begin{array}{cc} -13.1716 & 0 \\ 0 & -18.8284 \end{array}$$

```
VB=
```

$$\begin{array}{cc} 0.7071 & -0.3162 \\ 0.7071 & 0.9487 \end{array}$$

DB =

$$\begin{array}{cc} 3 & 0 \\ 0 & -1 \end{array}$$

As we can see, the eigenvalues for A are −13.17 and −18.82, and the eigenvectors are

$$\begin{bmatrix} .97 \\ .2 \end{bmatrix} \text{ and } \begin{bmatrix} -.63 \\ .77 \end{bmatrix}.$$

EOCE 6.3

Consider matrix B as given in EOCE 6.2.
    Find e^{Bt}.

SOLUTION

B is not diagonal. As discussed in this chapter, we need to find the matrix P that contains the eigenvectors of B, then we will have

$$e^{Bt} = Pe^{P^{-1}BPt}P^{-1}$$

We can use MATLAB to find $P^{-1}BP$, as in the EOCE6_3 script with the output

VB=

$$\begin{array}{cc} 0.7071 & -0.3162 \\ 0.7071 & 0.9487 \end{array}$$

DB=

$$\begin{array}{cc} 5 & 0 \\ 0 & 1 \end{array}$$

Inverse_of_P=

$$\begin{array}{cc} 1.0607 & 0.3536 \\ -0.7906 & 0.7906 \end{array}$$

Diagonal_form_of_B=

$$\begin{array}{cc} 5 & 0 \\ 0 & 1 \end{array}$$

We can see that the diagonal form of B is

$$P^{-1}BP = \begin{bmatrix} 5 & 0 \\ 0 & 1 \end{bmatrix},$$

and

$$e^{P^{-1}BPt} = \begin{bmatrix} e^{5t} & 0 \\ 0 & e^{t} \end{bmatrix}$$

$$e^{Bt} = Pe^{P^{-1}BPt}P^{-1} = \begin{bmatrix} .7071 & -.3162 \\ .7071 & .9487 \end{bmatrix} \begin{bmatrix} e^{5t} & 0 \\ 0 & e^{t} \end{bmatrix} \begin{bmatrix} 1.0607 & .3536 \\ -.7906 & .7906 \end{bmatrix}$$

$$e^{Bt} = \begin{bmatrix} .7071e^{5t} & -.3162e^{t} \\ 7071e^{5t} & .9487e^{t} \end{bmatrix} \begin{bmatrix} 1.0607 & .3536 \\ -.7906 & .7906 \end{bmatrix}.$$

And finally,

$$e^{Bt} = \begin{bmatrix} .75e^{5t}+.25e^{t} & .25e^{5t}-.25e^{t} \\ .75e^{5t}-.75e^{t} & .25e^{5t}+.75e^{t} \end{bmatrix}.$$

EOCE 6.4

Consider the following state equations.

$$\frac{d}{dt}z_1 = z_2$$

$$\frac{d}{dt}z_2 = 6z_2 - 5z_1.$$

We can see that these state equations are a result of a second-order system with zero input, since no inputs appear to the right side of the two equations.

Assume that the initial conditions are $z_1(0) = 1$ and $z_2(0) = -1$. What are the states of this system?

SOLUTION

1. Using the Laplace domain
   The general state-space solution in the Laplace domain is

$$\underline{Z}(s) = (SI - A)^{-1}\underline{z}(0) + (sI - A)^{-1}B\underline{X}(s),$$

where A is the dynamic matrix of the system, B is the input matrix ($B = [0]$, no inputs), X is the input vector, and Z is the state vector. In this example

$$A = \begin{bmatrix} 0 & 1 \\ -5 & 6 \end{bmatrix} \text{ and } B = \begin{bmatrix} 0 \\ 0 \end{bmatrix},$$

with the initial conditions

$$\underline{z}(0) = \begin{bmatrix} 1 \\ -1 \end{bmatrix}.$$

Substituting these values in the state equation we get

$$\underline{Z}(s) = \frac{1}{s^2 - 6s + 5} \begin{bmatrix} s-6 & 1 \\ -5 & s \end{bmatrix} \begin{bmatrix} 1 \\ -1 \end{bmatrix} = \begin{bmatrix} \dfrac{s-7}{s^2 - 6s + 5} \\ \dfrac{-5-s}{s^2 - 6s + 5} \end{bmatrix}.$$

Finally, the states are

$$\underline{Z}(s) = \begin{bmatrix} \dfrac{-1/2}{s-5} + \dfrac{3/2}{s-1} \\ \dfrac{-5/2}{s-5} + \dfrac{3/2}{s-1} \end{bmatrix}.$$

If we take the inverse transform we get

$$\underline{z}(t) = \begin{bmatrix} z_1(t) \\ z_2(t) \end{bmatrix} = \begin{bmatrix} -0.5e^{5t} & + 1.5e^t \\ -2.5e^{5t} & + 1.5e^t \end{bmatrix} t \geq 0.$$

2. Using the time domain

Since A is not in the diagonal form we need to use the following form of the solution

$$\underline{w}(t) = \exp\left[P^{-1}APt\right]\underline{w}(0) + \int_0^t \exp\left[P^{-1}AP\,(t-\tau)\right]P^{-1}Bx(\tau)d\tau,$$

where $\underline{z}(t) = P\underline{w}(t)$ and $\underline{w}(0) = P^{-1}\underline{z}(0)$. Since B is a zero matrix

$$\underline{z}(t) = P\exp\left[P^{-1}APt\right]P^{-1}\underline{z}(0).$$

To solve for $\underline{z}(t)$ we need to find P, P^{-1}, and $P^{-1}AP$. We do that using the MATLAB script EOCE6_4.

The output is

P =

$$\begin{bmatrix} -0.7071 & -0.1961 \\ -0.7071 & -0.9806 \end{bmatrix}$$

Pinverse =

$$\begin{bmatrix} -1.7678 & 0.3536 \\ 1.2748 & -1.2748 \end{bmatrix}$$

PinverseAP =

$$\begin{bmatrix} 1.0000 & 0.0000 \\ 0 & 5.0000 \end{bmatrix}$$

Therefore, the solution for the states is

$$\underline{Z}(t) = \begin{bmatrix} -0.7071 & -0.1961 \\ -0.7071 & -0.9806 \end{bmatrix} \begin{bmatrix} e^t & 0 \\ 0 & e^{5t} \end{bmatrix} \begin{bmatrix} -2.1213 \\ 2.5495 \end{bmatrix}$$

$$\underline{Z}(t) = \begin{bmatrix} -0.7071 & -0.1961 \\ -0.7071 & -0.9806 \end{bmatrix} \begin{bmatrix} -2.1213e^t \\ 2.5495e^{5t} \end{bmatrix} \quad t \geq 0.$$

Finally,

$$\begin{bmatrix} z_1(t) \\ z_2(t) \end{bmatrix} = \begin{bmatrix} 1.5e^t & - & 0.5e^{5t} \\ 1.5e^t & - & 2.5e^{5t} \end{bmatrix} \quad t \geq 0.$$

EOCE 6.5

Consider the mechanical system in Figure 6.5.

FIGURE 6.5 System for EOCE 6.5.

The rod to the left of the mass M can be modeled by a translational spring with spring constant K, in parallel with a translational damper of constant B. Assume the mass moves freely. As seen before, the differential equation representing the system is

$$M\frac{d^2}{dt}y(t) + B\frac{d}{dt}y(t) + Ky(t) = x(t).$$

In a general form we can write the above equation as

$$\frac{d^2}{dt}y(t) + 2\xi w_n\frac{d}{dt}y(t) + w_n^2 y(t) = \frac{x(t)}{M}.$$

The characteristic equation is

$$m^2 + 2\xi w_n m + w_n^2 = 0,$$

with roots at

$$m_{1,2} = -\xi w_n \pm w_n\sqrt{\xi^2 - 1}.$$

We call ξ the damping ratio, where

$$\xi = \frac{B}{2\sqrt{KM}}.$$

In our example

$$w_n = \sqrt{\frac{K}{M}}.$$

We have critical damping if $\xi = 1$, underdamping if ξ is less than one and greater than zero, and overdamping if $\xi > 1$.

Let $M = 1$ kg and $K = 40$ newton/meter.

For critical damping, $\xi = 1$, and $w_n = \sqrt{\frac{K}{M}} = 6.3246$.

For an underdamped case, let $\xi = 0.5$, while keeping the same value for w_n.
For the overdamped case let $\xi = 2$, while keeping the same value for w_n as well.

Assume zero initial conditions and plot the displacement $y(t)$ vs. time for all cases above if $x(t)$ is a constant force of 1 newton.

SOLUTION

We will use state equations to solve this problem using MATLAB.

Let $z_1(t) = y(t)$ and $z_2(t) = \frac{d}{dt}y(t)$ to get the state equations as

$$\frac{d}{dt}z(t) = \begin{bmatrix} 0 & 1 \\ -w_n^2 & -2\xi w_n \end{bmatrix} z(t) + \begin{bmatrix} 0 \\ \frac{1}{M} \end{bmatrix} x(t),$$

with the matrices

$$A = \begin{bmatrix} 0 & 1 \\ -w_n^2 & -2\xi w_n \end{bmatrix} \text{ and } B = \begin{bmatrix} 0 \\ \dfrac{1}{M} \end{bmatrix},$$

and the initial conditions

$$\underline{z}(0) = \begin{bmatrix} 0 \\ 0 \end{bmatrix}.$$

A MATLAB function will be used to define the state equation as eoce5.m.

Then the script EOCE6_5 is used that we will call the "ode23" function, which is a MATLAB function with initial conditions and initial and final values for the time needed for the simulation.

The plots are given in Figure 6.6.

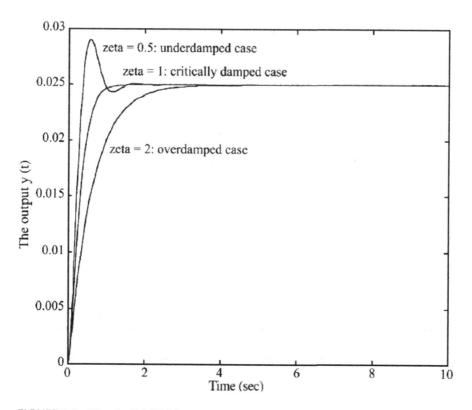

FIGURE 6.6 Plots for EOCE 6.5.

EOCE 6.6

Consider the system in Figure 6.7.

The rod to the left of the mass M_1 can be modeled by a translational spring with spring constant k_1. The rod to the right of the mass M_1 can be modeled by a translational spring with spring constant k_2. The rod to the right of the mass M_2 can be modeled by a translational spring with spring constant k_3.

Find the displacements $y_1(t)$ and $y_2(t)$ for different initial conditions.

SOLUTION

The system has two degrees of freedom (movement of M_1 and M_2) and it will have two coupled second-order differential equations. Each differential equation can be represented as two first-order differential equations. Hence, the system given is a system with four state variables. We can use Newton's law to write the two coupled differential equations as

$$M_1 \frac{d^2}{dt} y_1 + k_1 y_1 + k_2 y_1 - k_2 y_2 = 0$$

$$M_2 \frac{d^2}{dt} y_2 - k_2 y_1 + k_2 y_2 + k_3 y_2 = 0.$$

If we let all the spring constants be 1 and the values of the masses be 1, we will have

$$\frac{d^2}{dt} y_1 + y_1 + y_1 - y_2 = 0$$

$$\frac{d^2}{dt} y_2 - y_1 + y_2 + y_2 = 0.$$

If we now let

$$z_1 = y_1$$

$$z_2 = y_2$$

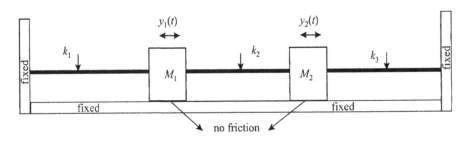

FIGURE 6.7   System for EOCE 6.6.

and

$$z_3 = \frac{d}{dt} y_1$$

$$z_4 = \frac{d}{dt} y_2$$

then the following state equation is obtained

$$\frac{d}{dt} \underline{z} = \begin{bmatrix} \dfrac{d}{dt} z_1 \\[2mm] \dfrac{d}{dt} z_2 \\[2mm] \dfrac{d}{dt} z_3 \\[2mm] \dfrac{d}{dt} z_4 \end{bmatrix} = \begin{bmatrix} 0 & 0 & 1 & 0 \\ 0 & 0 & 0 & 1 \\ -2 & 1 & 0 & 0 \\ 1 & -2 & 0 & 0 \end{bmatrix} \underline{z}.$$

We will let the system vibrate at its natural frequencies with specified initial conditions. The natural frequencies can be calculated by using the MATLAB command to find the eigenvalues of the system as

```
A= [0 0 1 0; 0 0 0 1; -2 1 0 0;1 -2 0 0];
poles=eig (A)
```

with the output

```
poles =
0.0000+ 1.7321i
0.0000- 1.7321i
0.0000+ 1.0000i
0.0000- 1.0000i
```

Thus, the natural frequencies are at 1 and 1.7321 rad/sec. We will use MATLAB to study the response of the system for different initial conditions. We will start writing the function that defines the system as eoce6.m. Next, we write the script EOCE6_6 that will call the MATLAB "ode23" function and the function that defines the system. Since we will have different initial conditions, we will write a very general script that will communicate with the user.

We will first run the script above and let the masses M_1 and M_2 move two units to the right. In this case, the motions of M_1 and M_2 are in-phase. The following is a sample dialogue obtained by typing EOCE6_6 at the MATLAB prompt

```
Initial time for simulation = 0
t_initial =
0
Final time for simulation = 10
```

```
t_final =
10
[z1(t_initial) z2(t_initial) dz1(t_initial) dz2
(t_initial)] = [2 2 0 0]
z0 =
2 2 0 0
Title = The two movements are in-phase: M1 and M2 are
moved each 2 units to the right.
plot_title =
The two movements are in-phase: M1 and M2 are moved each
2 units to the right.
```

The plots are shown in Figure 6.8.

In this first case, the central spring with the spring constant k_2 is not stretched or compressed. Since the initial displacements are equal, the output displacements are in-phase and the fourth-order system behaves as a second-order system.

Now let us try moving M_1 two units to the right and moving M_2 two units to the left at the same time and then release the masses. This is done by setting $z_1(0) = 2$ and $z_2(0) = -2$.

This is another dialogue with MATLAB and we initiate it by typing EOCE6_6 at the MATLAB prompt.

```
Initial time for simulation = 0
t_Initial =
0
Final time for simulation = 10
t_final =
```

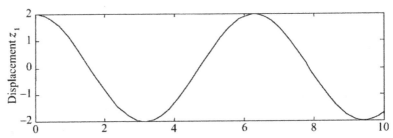

The two movements are in-phase: M_1 and M_2 are moved each two units to the right

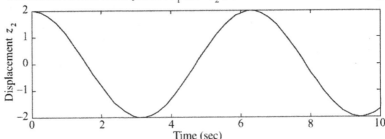

Time (sec)

FIGURE 6.8 Plots for EOCE 6.6.

```
10
z1(t_initial) z2 (t_initial) dz1 (t_initial) dz2
(t_initial)] = [2 -2 0 0]
z0 =
2 -2 0 0
Title = The out-of-phase movements: M1 two units right,
M2 two units left.
plot_title =
The out-of-phase movements: M1 two units right, M2 two
units left.
```

The plots are shown in Figure 6.9.

In this second case, the central spring is compressed and all spring constants are involved. This example is constructed for equal k and M values. You may experiment with the functions presented using various values for the masses, the spring constants, and the initial conditions.

EOCE 6.7

Consider the system in Figure 6.10.

The rod to the left of the mass M can be modeled by a translational spring with spring constant k_1, in parallel with a translational damper of constant B_1. The rod to the right of the mass M can be modeled by a translational spring with spring constant k_2, in parallel with a translational damper of constant B_2.

Find the displacement $y(t)$ due to an input $x(t)$.

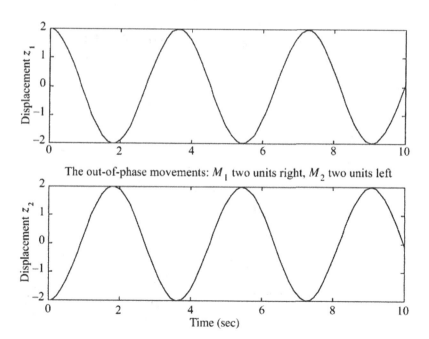

FIGURE 6.9 Plots for EOCE 6.6.

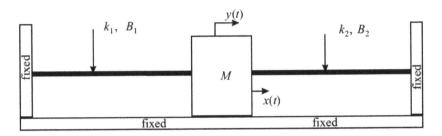

FIGURE 6.10   System for EOCE 6.7.

SOLUTION

The free-body diagram that describes the system is shown in Figure 6.11, where $x(t)$ is the input displacement and $y(t)$ is the output displacement. k_1 and k_2 are the stiffness elements and B_1 and B_2 are the friction elements. The differential equation describing the system is

$$M\frac{d^2}{dt}y + (B_1 + B_2)\frac{d}{dt}y + (k1 + k2)y = x.$$

If we let $k = k_1 + k_2$ and $B = B_1 + B_2$, the differential equation becomes

$$M\frac{d^2}{dt}x + B\frac{d}{dt}x + kx = x.$$

This system is similar to the system given previously in EOCE 6.5. Let $M = 1$, $k = 40$, and $B = 12$. Let $z_1(t) = y(t)$ and $z_2(t) = \frac{d}{dt}y(t)$. Therefore, the state equation is

$$\frac{d}{dt}\underline{z}(t) = \begin{bmatrix} 0 & 1 \\ -40 & -12 \end{bmatrix}\underline{z}(t) + \begin{bmatrix} 0 \\ 1 \end{bmatrix}\underline{x}(t),$$

and the output equation is

$$y = \begin{bmatrix} 1 & 0 \end{bmatrix}\begin{bmatrix} z_1 \\ z_2 \end{bmatrix} + [0]x,$$

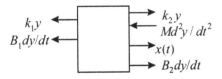

FIGURE 6.11   Free-body diagram for EOCE 6.7.

with the matrices

$$A = \begin{bmatrix} 0 & 1 \\ -40 & -12 \end{bmatrix}, B = \begin{bmatrix} 0 \\ 1 \end{bmatrix}, C = \begin{bmatrix} 1 & 0 \end{bmatrix}, \text{ and } D = \begin{bmatrix} 0 \end{bmatrix},$$

and the initial condition vector

$$\underline{z}(0) = \begin{bmatrix} 0 \\ 0 \end{bmatrix},$$

where we assumed that the system is relaxed (zero initial conditions). The solution for the state equation is

$$\underline{z}(t) = \int_0^t \exp\left[A(t-\tau)\right] Bx(\tau) d\tau,$$

where e^{At} is the inverse Laplace transform of $(sI - A)^{-1}$.

$$(sI - A)^{-1} = \begin{bmatrix} s & -1 \\ 40 & s+12 \end{bmatrix} = \frac{1}{(s^2 + 12s + 40)} \begin{bmatrix} s+12 & 1 \\ -40 & s \end{bmatrix}$$

$$(sI - A)^{-1} = \begin{bmatrix} \dfrac{s+12}{(s^2 + 12s + 40)} & \dfrac{1}{(s^2 + 12s + 40)} \\ \dfrac{-40}{(s^2 + 12s + 40)} & \dfrac{s}{(s^2 + 12s + 40)} \end{bmatrix}.$$

Finally,

$$(sI - A)^{-1} = \begin{bmatrix} \dfrac{s+6+6}{(s+6)^2 + 2^2} & \dfrac{1}{(s+6)^2 + 2^2} \\ \dfrac{-40}{(s+6)^2 + 2^2} & \dfrac{s+6-6}{(s+6)^2 + 2^2} \end{bmatrix}.$$

The inverse transform of the last matrix is

$$\begin{bmatrix} e^{-6t}\cos(2t) + 3e^{-6t}\sin(2t) & 1/2\left(e^{-6t}\sin(2t)\right) \\ -20e^{-6t}\sin(2t) & e^{-6t}\cos(2t) - 3e^{-6t}\sin(2t) \end{bmatrix} = e^{At} = \Phi(t).$$

Therefore, using the transition matrix we can write the solution for the state vector as

$$z(t) = \int_0^t \exp\left[A(t-\tau)\right]Bx(\tau)d\tau$$

$$z(t) = \int_0^t \begin{bmatrix} e^{-6(t-\tau)}\cos\left(2(t-\tau)\right) + 3e^{-6(t-\tau)}\sin\left(2(t-\tau)\right) \\ -20e^{-6(t-\tau)}\sin\left(2(t-\tau)\right) \end{bmatrix} \delta(\tau)d\tau,$$

and finally

$$z(t) = \begin{bmatrix} e^{-6t}\cos(2t) + 3e^{-6t}\sin(2t) \\ -20e^{-6t}\sin(2t) \end{bmatrix}.$$

The last result was obtained using the sifting property of the impulse function. The output equation $y(t)$ is

$$y = \begin{bmatrix} 1 & 0 \end{bmatrix}\begin{bmatrix} z_1 \\ z_2 \end{bmatrix} = \left[e^{-6t}\cos(2t) + 3e^{-6t}\sin(2t)\right]u(t).$$

We can see that the system is stable because the real part of the poles is negative at −6. We can use MATLAB to find the impulse response of the system and compare it with the analytical solution that we just found. To do that we write the script EOCE6_7. The output is shown in Figure 6.12.

FIGURE 6.12   Plots for EOCE 6.7.

FIGURE 6.13   System for EOCE 6.8.

EOCE 6.8

Consider the system shown in Figure 6.13.

Where $\theta(t)$ is the rotational angle, J represents the moment of inertia, where the mass of the rod is concentrated, k is the stiffness coefficient, w is the angular velocity, and τ is the torque applied. Assume the system was relaxed before applying the torque. Find the rotational angle resulting from the applied torque.

SOLUTION

The free-body diagram is shown in Figure 6.14.

The differential equation describing this rotational system is

$$J\frac{d}{dt}w(t) + k\theta(t) = \tau(t).$$

Rewriting as a function of $\theta(t)$, the equation becomes

$$J\frac{d^2}{dt}\theta(t) + k\theta(t) = \tau(t).$$

This is a second-order differential equation in $\theta(t)$, where $\tau(t)$ is the input torque.

If we let $z_1 = \theta(t)$, and $z_2 = \frac{d}{dt}\theta(t)$, then we can write the system in state-space as

$$\frac{d}{dt}\underline{z}(t) = \begin{bmatrix} 0 & 1 \\ -k/J & 0 \end{bmatrix}\underline{z}(t) + \begin{bmatrix} 0 \\ 1/J \end{bmatrix}\underline{x}(t),$$

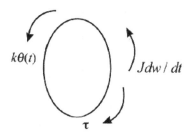

FIGURE 6.14   Free-body diagram for EOCE 6.8.

and the output equation as

$$y = \begin{bmatrix} 1 & 0 \end{bmatrix} \begin{bmatrix} z_1 \\ z_1 \end{bmatrix} + [0]x.$$

The initial condition vector is

$$\underline{z}(0) = \begin{bmatrix} 0 \\ 0 \end{bmatrix}.$$

If the input torque is an impulse of unity strength, $J = 1$ kg.m², and $k = 1$ newton. meter/rad, we can find the displacement angle $\theta(t)$.

For the given values the system state equation becomes

$$\frac{d}{dt}\underline{z}(t) = \begin{bmatrix} 0 & 1 \\ -1 & 0 \end{bmatrix} \underline{z}(t) + \begin{bmatrix} 0 \\ 1 \end{bmatrix} x(t),$$

and the output equation becomes

$$y = \begin{bmatrix} 1 & 0 \end{bmatrix} \begin{bmatrix} z_1 \\ z_2 \end{bmatrix} + [0]x,$$

with the initial conditions

$$\underline{z}(0) = \begin{bmatrix} 0 \\ 0 \end{bmatrix}.$$

e^{At} is the inverse Laplace transform of $(sI - A)^{-1}$ and is calculated next.

$$(sI - A)^{-1} = \begin{bmatrix} s & -1 \\ 1 & s \end{bmatrix}^{-1} = \frac{1}{(s^2+1)}\begin{bmatrix} s & 1 \\ -1 & s \end{bmatrix}$$

$$(sI - A)^{-1} = \begin{bmatrix} \dfrac{s}{(s^2+1)} & \dfrac{1}{(s^2+1)} \\ \dfrac{-1}{(s^2+1)} & \dfrac{s}{(s^2+1)} \end{bmatrix}$$

$$e^{At} = \Phi(t) = \begin{bmatrix} \cos(t) & \sin(t) \\ -\sin(t) & \cos(t) \end{bmatrix}.$$

The solution for the states now is calculated as

$$\underline{z}(t) = \int_0^t \exp\left[A(t - \tau) Bx(\tau) d\tau \right].$$

With $\Phi(t)$ as calculated above

$$\underline{z}(t) = \int_0^t \begin{bmatrix} \cos(t-\tau) & \sin(t-\tau) \\ -\sin(t-\tau) & \cos(t-\tau) \end{bmatrix} \begin{bmatrix} 0 \\ 1 \end{bmatrix} \delta(\tau) d\tau$$

$$\underline{z}(t) = \int_0^t \begin{bmatrix} \sin(t-\tau) \\ \cos(t-\tau) \end{bmatrix} \delta(\tau) d\tau.$$

Using the sifting property we have the displacement and the velocity angles as

$$\underline{z}(t) = \begin{bmatrix} \theta(t) \\ \dfrac{d}{dt}\theta(t) \end{bmatrix} = \begin{bmatrix} \sin(t) \\ \cos(t) \end{bmatrix}.$$

We can use MATLAB to display the angular displacement and the angular velocity by writing the script EOCE6_8.

The output plot is shown in Figure 6.15.

FIGURE 6.15 Plots for EOCE 6.8.

EOCE 6.9

Consider the thermal system shown in Figure 6.16, where $x(t)$ is the rate at which heat is supplied to the left capacitor. R_1 and R_2 are the thermal resistors, T_1 and T_2 are the temperatures in the two sections as shown, and ambient is the outside temperature. Assume, excluding R_1 and R_2, that the system is insulated.

Let T_1 and T_2 be the two state variables. Find T_1 and T_2.

SOLUTION

Using the general relation

$$\frac{d}{dt}T = \frac{1}{C}(rate\_heat\_in - rate\_heat\_out),$$

we can write

$$\frac{d}{dt}T_1 = \frac{1}{C_1}\left[x(t) - \frac{1}{R_1}(T_1 - T_2)\right]$$

and

$$\frac{d}{dt}T_2 = \frac{1}{C_2}\left[\frac{1}{R_1}(T_1 - T_2) - \frac{1}{R_2}(T_2 - T)\right].$$

As you may notice, this system has two inputs, T and $x(t)$.

Let $z_1 = T_1$ and $z_2 = T_2$. Then the state equation is

$$\frac{d}{dt}\underline{z}(t) = \begin{bmatrix} \dfrac{-1}{C_1R_1} & \dfrac{1}{C_1R_1} \\ \dfrac{1}{C_1R_1} & \dfrac{-(R_2+R_1)}{C_2R_2R_1} \end{bmatrix} \underline{z}(t) + \begin{bmatrix} 1/C_1 & 0 \\ 0 & 1/R_2C_2 \end{bmatrix}\begin{bmatrix} x(t) \\ T \end{bmatrix},$$

and the output equation is

$$y = \begin{bmatrix} 1 & 0 \end{bmatrix}\begin{bmatrix} z_1 \\ z_2 \end{bmatrix} + [0]x,$$

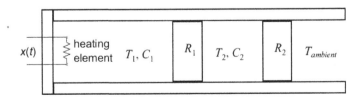

FIGURE 6.16 System for EOCE 6.9.

with the initial condition

$$\underline{z}(0) = \begin{bmatrix} z_1(0) \\ z_2(0) \end{bmatrix}.$$

For the purpose of simulation only, let $R_1 = R_2 = 1$ ohm, $T = 104$ kelvin, $C_1 = C_2 = 1$, and $x(t) = 0$. Assume zero initial conditions. The state equations become

$$\frac{d}{dt}\underline{z}(t) = \begin{bmatrix} -1 & 1 \\ 1 & -2 \end{bmatrix}\underline{z}(t) + \begin{bmatrix} 0 \\ 104 \end{bmatrix},$$

and the output becomes

$$y = \begin{bmatrix} 1 & 0 \end{bmatrix}\begin{bmatrix} z_1 \\ z_2 \end{bmatrix} + [0]x,$$

with the initial conditions

$$\underline{z}(0) = \begin{bmatrix} 0 \\ 0 \end{bmatrix}.$$

The eigenvalues of this system are found by setting (det $(sI - A)) = 0$), or by using MATLAB as in the following.

```
A= [-1 1 ; 1 -2];
eig(A)
```

The output is then

```
- 0.3820

- 2.6180
```

e^{At} is the inverse Laplace transform of $(sI - A)^{-1}$ and is needed to find the solution of the state equation.

$$(sI - A)^{-1} = \begin{bmatrix} s+1 & -1 \\ -1 & s+2 \end{bmatrix}^{-1} = \frac{1}{(s^2+3s+1)}\begin{bmatrix} s+2 & 1 \\ 1 & s+1 \end{bmatrix}$$

$$(sI - A)^{-1} = \begin{bmatrix} \dfrac{s+2}{(s^2+3s+1)} & \dfrac{1}{(s^2+3s+1)} \\ \dfrac{1}{(s^2+3s+1)} & \dfrac{s+1}{(s^2+4s+1)} \end{bmatrix}$$

Using partial fraction expansion we get

$$(sI - A)^{-1} = \begin{bmatrix} \dfrac{0.2764}{s+2.618} + \dfrac{0.7236}{s+0.382} & \dfrac{-0.4472}{s+2.618} + \dfrac{0.4472}{s+0.382} \\[2mm] \dfrac{-0.4472}{s+2.618} + \dfrac{0.4472}{s+0.382} & \dfrac{0.7236}{s+2.618} + \dfrac{0.2764}{s+0.382} \end{bmatrix}.$$

The inverse transform of the last matrix is the state-transition matrix.

$$\begin{bmatrix} 0.2764e^{-2.618t} + 0.7236e^{-0.382t} & -4472e^{-2.618t} + 0.4472e^{-0.382t} \\[2mm] -0.4472e^{-2.618t} + 0.4472e^{-0.382t} & 0.7236e^{-2.618t} + 0.274e^{-0.382t} \end{bmatrix} = e^{At} = \Phi(t)$$

The state solution is

$$\underline{z}(t) = \int_0^t \exp\left[A(t-\tau)\right] Bx(\tau)\, d\tau.$$

With $\Phi(t)$ substituted in the above convolution integral we have

$$\underline{z}(t) = \int_t^t \begin{bmatrix} 104\left(-0.4472e^{-2.618(t-\tau)} + 0.4472e^{-0.382(t-\tau)}\right) \\[2mm] 104\left(0.7236e^{-2.618(t-\tau)} + 0.2764e^{-0.382(t-\tau)}\right) \end{bmatrix} d\tau$$

$$\underline{z}(t) = \begin{bmatrix} 104\left[0.1708\left(e^{-2.618t}-1\right)-1.1707\left(e^{-0.382t}-1\right)\right] \\[2mm] 104\left[-0.2764\left(e^{-2.618t}-1\right)-0.7236\left(e^{-0.382t-1}\right)\right] \end{bmatrix},$$

where

$$z_1(t) = T_1(t) = 104\left(0.1708\left(e^{-2.618t}-1\right)-1.1707\left(e^{-0.382t}-1\right)\right).$$

We can plot the temperature in the first capacitor using MATLAB by writing the script EOCE6_9.

The output is shown in Figure 6.17.

EOCE 6.10

Consider the circuit in Figure 6.18.

Find the voltages in the capacitors and the resistors as well as the current in the inductor with $x(t)$ as a step input of unity magnitude. Use MATLAB to verify the results. What are the poles of the system? What can you say about its stability?

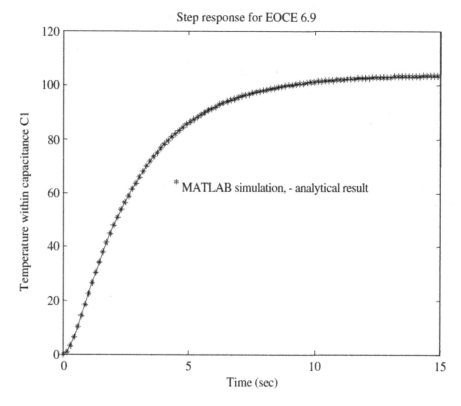

FIGURE 6.17 Step response for EOCE 6.9.

FIGURE 6.18 Circuit for EOCE 6.10.

SOLUTION

We will use the state-space approach and let the voltages in the capacitors be represented by the states z_1 and z_2. We will also let the third state be the current in the inductor and call it z_3. The voltages in the resistors will be calculated after we find the three independent states. Since our system has three independent energy storage elements, we will only have these three states. Remember that resistors are not storage elements to start with.

Using nodal analysis as well as some loop equations, we come up with the following equations:

$$\frac{d}{dt}z_1 + \frac{d}{dt}z_2 - z_3 = 0$$

$$z_1 = z_2 + \frac{d}{dt}z_2$$

$$x = 10z_3 + \frac{d}{dt}z_3 + z_1.$$

We can rearrange the equations above to get

$$\frac{d}{dt}z_1 = -z_1 + z_2 + z_3$$

$$\frac{d}{dt}z_2 = z_1 - z_2$$

$$\frac{d}{dt}z_3 = -z_1 - 10z_3 + x.$$

In the state-space form we write the above system as

$$\begin{bmatrix} \dfrac{d}{dt}z_1 \\ \dfrac{d}{dt}z_2 \\ \dfrac{d}{dt}z_3 \end{bmatrix} = \begin{bmatrix} -1 & 1 & 1 \\ 1 & -1 & 0 \\ -1 & 0 & -10 \end{bmatrix} \begin{bmatrix} z_1 \\ z_2 \\ z_3 \end{bmatrix} + \begin{bmatrix} 0 \\ 0 \\ 1 \end{bmatrix} x,$$

where

$$A = \begin{bmatrix} -1 & 1 & 1 \\ 1 & -1 & 0 \\ -1 & 0 & -10 \end{bmatrix}$$

and

$$B = \begin{bmatrix} 0 \\ 0 \\ 1 \end{bmatrix}.$$

Since A is not in the diagonal form, we need to diagonalize it in order to arrive at the analytical solution for the states.

The real-time solution in this case is

$$\underline{w}(t) = \exp\left[P^{-1}APt\right]\underline{w}(0) + \int_0^t \exp\left[P^{-1}AP(t-\tau)\right]P^{-1}Bx(\tau)d\tau,$$

with $\underline{z}(t) = P\underline{w}(t)$, where P is the matrix that diagonalizes A. Since $\underline{w}(0) = P^{-1}\underline{z}(0)$, and because $\underline{z}(0) = \underline{0}$, and $\underline{w}(0) = \underline{0}$, we get

$$\underline{w}(t) = \int_0^t \exp\left[P^{-1}AP(t-\tau)\right]P^{-1}Bx(\tau)d\tau.$$

We will start by finding the matrix P that diagonalizes A. P is the matrix that contains the eigenvectors of A. We will use MATLAB to find the eigenvalues and eigenvectors of A. Knowing the eigenvalues of A, we can determine the stability of the system. If the real parts of the eigenvalues are all negative, then the system is stable. We will write the following script in MATLAB to do that.

```
A = [-1 1 1 ; 1 -1 0 ; -1 0 -10] ;
[VA, DA] = eig (A);
P = VA
Pinverse = inv (P)
PinverseAP = inv (P) *A*P
```

The output is

```
P =
```

−0.6875	−0.7259	−0.1132
−0.7229	0.6816	0.0127
0.0691	0.0915	0.9935

```
Pinverse =
```

−0.6941	−0.7299	−0.0698
−0.7383	0.6932	−0.0930
0.1163	−0.0131	1.0200

```
PinverseAP =
```

−0.0490	0.0000	0.0000
0.0000	−2.0650	0.0000
0.0000	0.0000	−9.8860

Since the eigenvalues are all negative, the system is stable. To find $\underline{w}(t)$, we need the following matrices first.

$$P^{-1}B = \begin{bmatrix} -0.0698 \\ -0.0930 \\ 1.0200 \end{bmatrix}$$

$$e^{P^{-1}APt} = \begin{bmatrix} e^{-.049t} & 0 & 0 \\ 0 & e^{-2.065t} & 0 \\ 0 & 0 & e^{-9.886t} \end{bmatrix}$$

$$e^{P^{-1}AP(t-\tau)}P^{-1}B = \begin{bmatrix} e^{-.049(t-\tau)} & 0 & 0 \\ 0 & e^{-2.065(t-\tau)} & 0 \\ 0 & 0 & e^{-9.886(t-\tau)} \end{bmatrix}\begin{bmatrix} -0.069 \\ -0.093 \\ 1.02 \end{bmatrix}$$

$$e^{P^{-1}AP(t-\tau)}P^{-1}B = \begin{bmatrix} -0.069e^{-.049(t-\tau)} \\ -0.093e^{-2.065(t-\tau)} \\ 1.02e^{-9.886(t-\tau)} \end{bmatrix}$$

The new state solution is

$$\underline{w}(t) = \int_0^t \begin{bmatrix} -0.069e^{-.049(t-\tau)} \\ -0.093e^{-2.065(t-\tau)} \\ 1.02e^{-9.886(t-\tau)} \end{bmatrix} d\tau = \begin{bmatrix} 1.408\left(e^{-0.049t}-1\right) \\ 0.045\left(e^{-2.065t}-1\right) \\ -0.1032\left(e^{-9.886t}-1\right) \end{bmatrix}.$$

With $\underline{z}(t) = P\underline{w}(t)$

$$\underline{z}(t) = \begin{vmatrix} \end{vmatrix}$$

$$\begin{bmatrix} -0.6875\left[1.408\left(e^{-0.049t}-1\right)\right]-0.7259\left[0.045\left(e^{-2.065t}-1\right)\right]-0.1132\left[-0.1032\left(e^{-9.886t}-1\right)\right] \\ -0.7229\left[1.408\left(e^{-0.049t}-1\right)\right]+0.6816\left[0.045\left(e^{-2.065t}-1\right)\right]+0.0127\left[-0.1032\left(e^{-9.886t}-1\right)\right] \\ 0.069\left[1.408\left(e^{-0.049t}-1\right)\right]+0.0915\left[0.045\left(e^{-2.065t}-1\right)\right]+0.9935\left[-0.1032\left(e^{-9.886t}-1\right)\right] \end{bmatrix},$$

or

$$\begin{bmatrix} z_1(t) \\ z_2(t) \\ z_3(t) \end{bmatrix} = \begin{bmatrix} -0.968e^{-0.049t}-0.0327e^{-2.065t}+0.0117e^{-9.886t}+1 \\ -1.0294e^{-0.049t}+0.0307e^{-2.065t}-0.0013e^{-9.886t}+1 \\ 0.0973e^{-0.049t}+0.0041e^{-2.065t}-0.1025e^{-9.886t} \end{bmatrix}.$$

Notice that as time becomes large, the two capacitors will have the same voltage. We will demonstrate that using plots. Now let us write the following MATLAB function that we will use to compare the output from MATLAB and the output using the results just obtained. We start with the function that defines the system and call it eoce10.m.

Next we write the script EOCE6_101 that calls the functions eoce10.m and ode23.

The plots are shown in Figure 6.19.

If we increase the simulation time we see that the voltages in the two capacitors approach the input step signal of unity magnitude, as shown in Figure 6.20.

Now we write the MATLAB script EOCE6_102 that simulates the hand calculations we obtained.

The plots are shown in Figure 6.21.

If we increase the simulation time to 100 seconds, we obtain the plots in Figure 6.22.

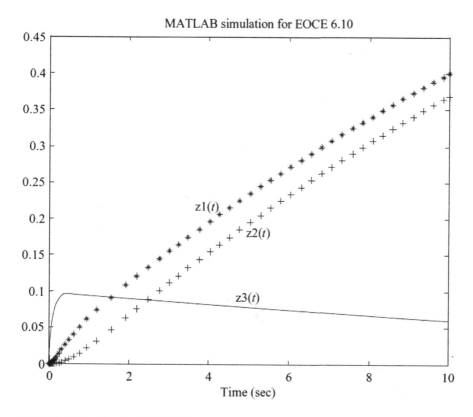

FIGURE 6.19   Plots for EOCE 6.10.

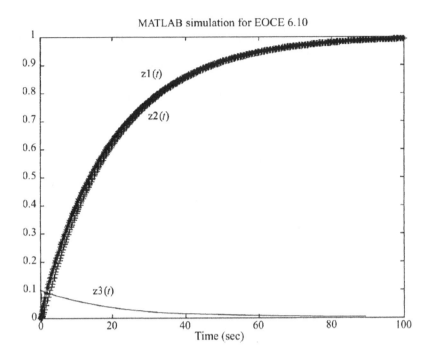

FIGURE 6.20 Plots for EOCE 6.10.

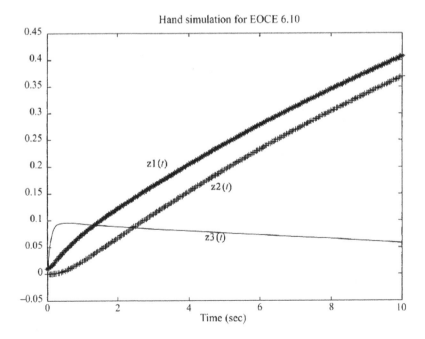

FIGURE 6.21 Plots for EOCE 6.10.

FIGURE 6.22   Plots for EOCE 6.10.

EOCE 6.11

Consider the following differential equation.

$$\frac{d^2}{dt^2}y(t) + 7\frac{d}{dt}y(t) + 12y(t) = 10x(t)$$

1. Find the impulse response representation.
2. Find the transfer function representation.
3. Find the block diagram representation.
4. Find the state-space representation
5. Find the step and the impulse response of the system.

SOLUTION

1. The impulse response representation
   We can calculate the impulse response in many ways.
   Graphically, the impulse response representation is calculated using the MATLAB script

```
num = [10];
den = [1 7 12] ;
impulse (num, den)
```

The output plot is shown in Figure 6.23.

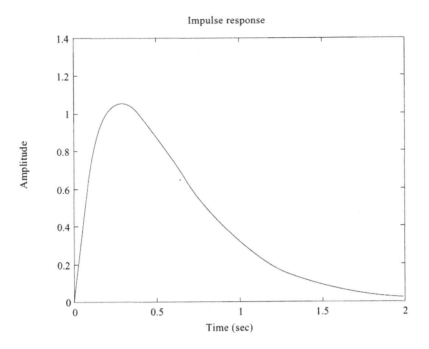

FIGURE 6.23 Plot for EOCE 6.11.

Using MATLAB symbolic toolbox function "dsolve," we can write the script EOCE6_111.
The output is

```
impulse =
-10/exp(t)^4 + 10/exp(t)^3
```

and the plot is shown in Figure 6.24.
Using partial fraction expansion with MATLAB, we write the script

```
num = [10];
den = [1 7 12] ;
[R, P, K] = residue (num, den)
```

The output is

```
R =
 - 10
10
P =
 - 4
 - 3
K =
 [ ]
```

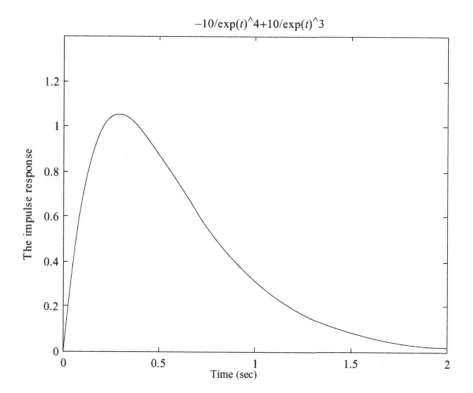

FIGURE 6.24 Plot for EOCE 6.11.

In this case

$$H(s) = \frac{-10}{s+4} + \frac{10}{s+3}$$

and

$$h(t) = \left(-10e^{-4t} + 10e^{-3t}\right)u(t)$$

which if plotted against time will produce the same figure as Figure 6.24.
2. The transfer function representation
   We can Laplace transform the differential equation with zero initial conditions to get

$$s^2Y(s) + 7sY(s) + 12Y(s) = 10X(s)$$

with the transfer function

$$H(s) = \frac{Y(s)}{X(s)} = \frac{10}{s^2 + 7s + 12}$$

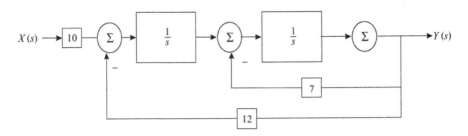

FIGURE 6.25   Block for EOCE 6.11.

3. The block diagram representation
   The block diagram can be drawn directly from $H(s)$ as shown in Figure 6.25.
4. The state-space representation
   We will let $z_1(t) = y(t)$ and $z_2(t) = \dfrac{d}{dt} y(t)$. Then

   $$\frac{d}{dt} z_1(t) = z_2(t)$$

   $$\frac{d}{dt} z_2(t) = -12z_1(t) - 7z_2(t) + 10x(t)$$

   and the state matrix equation is

   $$\frac{d}{dt} \underline{z} = A\underline{z} + B\underline{x}$$

   with

   $$A = \begin{bmatrix} 0 & 1 \\ -12 & -7 \end{bmatrix}, B = \begin{bmatrix} 0 \\ 10 \end{bmatrix}, \text{ and } \underline{z}(0) = \begin{bmatrix} 0 \\ 0 \end{bmatrix}$$

5. The step and the impulse responses
   To accomplish that we use the MATLAB script EOCE6_112. The output is shown in Figure 6.26.

EOCE 6.12

Consider the block diagram representation shown in Figure 6.27.
   Represent the system using the other representations discussed previously.

SOLUTION

1. The differential equation representation

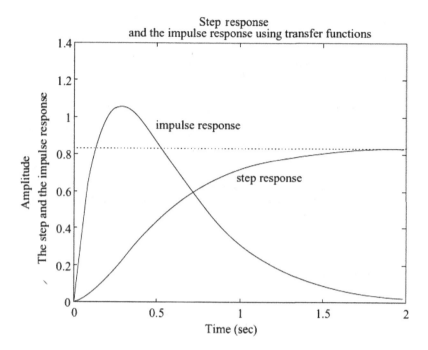

FIGURE 6.26 Plots for EOCE 6.11.

The output of the second integrator is y. Let the input to the second integrator be A. Let the output of the first integrator be B. The input to the first integrator is $(x - 12y)$. But

$$(x - 12y) = \frac{d}{dt}B$$

and

$$A = x + B = \frac{d}{dt}y$$

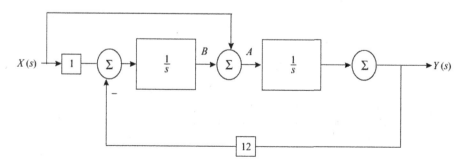

FIGURE 6.27 System for EOCE 6.12.

Also

$$\frac{d}{dt}A = \frac{d}{dt}x + \frac{d}{dt}B = \frac{d}{dt}x + x - 12y = \frac{d^2}{dt^2}y$$

The differential equation representing the system is then

$$\frac{d^2}{dt^2}y(t) + 12y(t) = x(t) + \frac{d}{dt}x(t)$$

2. The transfer function representation

Let us Laplace transform the differential equation with zero initial conditions to get

$$s^2Y(s) + 12Y(s) = X(s) + sX(s)$$

with the transfer function

$$\frac{Y(s)}{X(s)} = \frac{s+1}{s^2+12} = H(s)$$

Let us work with the block diagram directly. From the diagram

$$Y(s) = 1/s[A] = 1/s\left[X(s) + B\right]$$

Substituting for B we get

$$Y(s) = 1/s\left[X(s) + 1/s\left[X(s) - 12Y(s)\right]\right]$$

or

$$Y(s)\left[1 + \frac{12}{s^2}\right] = X(s)\left[\frac{1}{s} + \frac{1}{s^2}\right]$$

$$Y(s)\left[s^2 + 12\right] = X(s)[s+1]$$

The transfer function finally is

$$H(s) = \frac{Y(s)}{X(s)} = \frac{s+1}{s^2+12}$$

3. The impulse response representation

The impulse response representation can be calculated graphically and analytically. Graphically it can be found using the MATLAB command

```
Impulse ([1 1], [1 0 12]);
```

The plot is shown in Figure 6.28.
Analytically

$$H(s) = \frac{s+1}{s^2+12} = \frac{s}{s^2+12} + \frac{1}{s^2+12}$$

Using the table for the Laplace transform we can write $h(t)$ as

$$h(t) = \cos\left(\sqrt{12}t\right)u(t) + \frac{1}{\sqrt{12}}\sin\left(\sqrt{12}t\right)u(t)$$

This $h(t)$ agrees with Figure 6.28.
4. The state-space representation
   We will use MATLAB to find the state-space representation. Let us create the following script in MATLAB.

```
num = [1 1] ;
den = [1 0 12] ;
t = tf (num, den) % create the transfer function model
ssr = ss (t) % create the state-space representation
```

Impulse response

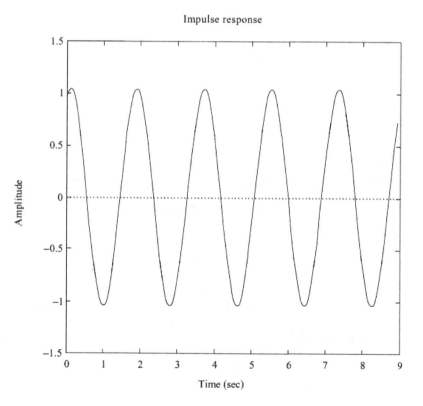

FIGURE 6.28 Plot for EOCE 6.12.

The output is
Transfer function

```
s + 1
--------
s^2 + 12
a =
```

```
              x1          x2
     x1        0     -3.00000
     x2   4.00000           0
```

```
b =
```

```
                   u1
      x1      1.00000
      x2            0
```

```
c =
```

```
              x1          x2
     y1   1.00000     0.25000
```

```
d =
```

```
             u1
      y1      0
```

This indicates that

$$A = \begin{bmatrix} 0 & -3 \\ 4 & 0 \end{bmatrix}, B = \begin{bmatrix} 1 \\ 0 \end{bmatrix}, C = \begin{bmatrix} 1 & 0.25 \end{bmatrix}, \text{ and } D = \begin{bmatrix} 0 \end{bmatrix}$$

If we do not use MATLAB and try to write the state-space representation, we will let $z_1(t) = y(t)$ and $z_2(t) = \dfrac{d}{dt}y(t) - x(t)$.
From the differential equations

$$\frac{d}{dt}z_1(t) = \frac{d}{dt}y(t) = z_2(t) + x(t)$$

$$\frac{d}{dt}z_2(t) = \frac{d^2}{dt}y(t) - \frac{d}{dt}x(t) = x(t) - 12z_1(t)$$

the state and the output matrix equations are

$$\frac{d}{dt}\underline{z} = A\underline{z} + B\underline{x}$$

$$\underline{y} = C\underline{z} + D\underline{x}$$

with

$$A = \begin{bmatrix} 0 & 1 \\ -12 & 0 \end{bmatrix}, \ B = \begin{bmatrix} 1 \\ 1 \end{bmatrix}, \underline{z}(0) = \begin{bmatrix} 0 \\ 0 \end{bmatrix},$$

$$C = \begin{bmatrix} 1 & 0 \end{bmatrix}, \text{ and } D = \begin{bmatrix} 0 \end{bmatrix}$$

with C as given, $y = z_1(t)$.

You also can see that the two state-space models or representations are different. Let us plot the output $y(t)$ in both cases and then we will make a conclusion.

For both models we first define the system in MATLAB by writing the function eoce12.m.

For the model we derived we write the MATLAB script EOCE6_121. The plot is shown in Figure 6.29.

For the MATLAB model the script EOCE6_122 is used The plot is shown in Figure 6.30.

It is clear that the two plots are identical and therefore, the state-space representations for the same system are not unique.

We can also obtain the state-space representation directly from the block diagram. Let the output of the second integrator be z_1 and the output of the first integrator be z_2.

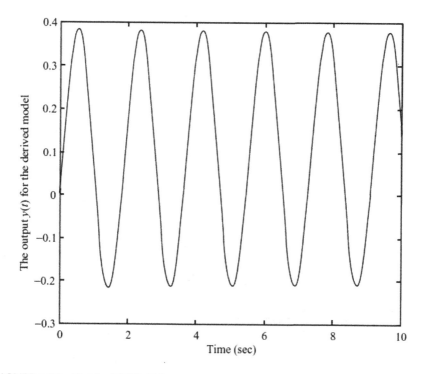

FIGURE 6.29 Plot for EOCE 6.12.

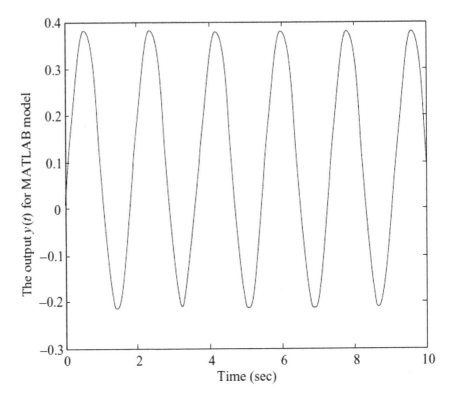

FIGURE 6.30 Plot for EOCE 6.12.

In this case

$$y = z_1$$

By differentiating we get

$$\frac{d}{dt}z_1 = z_2 + x$$

Also

$$\frac{d}{dt}z_2 = x - 12z_1$$

The resulting state matrices are

$$A = \begin{bmatrix} 0 & 1 \\ -12 & 0 \end{bmatrix}, B = \begin{bmatrix} 1 \\ 1 \end{bmatrix}, C = \begin{bmatrix} 1 & 0 \end{bmatrix}, \text{ and } D = \begin{bmatrix} 0 \end{bmatrix}$$

EOCE 6.13

Consider the impulse response of a certain system as

$$h(t) = 10e^{-2t}u(t) - 10e^{-4t}u(t)$$

Derive the other representations.

SOLUTION

1. The transfer function representation
   The inverse Laplace transform of the impulse response gives the transfer function. Laplace transform the given $h(t)$ to get

$$H(s) = \frac{10}{s+2} - \frac{10}{s+4}$$

or

$$H(s) = \frac{10(s+4) - 10(s+2)}{s^2 + 6s + 8} = \frac{20}{s^2 + 6s + 8}$$

2. The differential equation representation
   Remember that the impulse response is obtained by setting the initial conditions to zero.

$$H(s) = \frac{Y(s)}{X(s)} = \frac{20}{s^2 + 6s + 8}$$

or we can write

$$Y(s)\left[s^2 + 6s + 8\right] = X(s)[20]$$

Let us now inverse transform term by term to get

$$\frac{d^2}{dt}y(t) + 6\frac{d}{dt}y(t) + 8y(t) = 20x(t)$$

3. The block diagram representation
   The block diagram is shown in Figure 6.31.
4. The state-space representation
   First we will use MATLAB to do that. Here is the MATLAB script.

```
num = [20];
den = [ 1 6 8 ] ;
t = tf(num, den); % create the transfer function
model
ssr = ss(t) % create the state-space representation
```

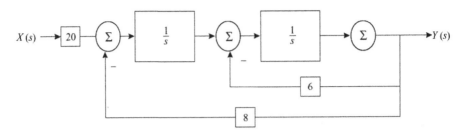

FIGURE 6.31 Block for EOCE 6.13.

The output is

a =

$$
\begin{array}{ccc}
 & \text{x1} & \text{x2} \\
\text{x1} & -6.00000 & -2.00000 \\
\text{x2} & 4.00000 & 0 \\
\end{array}
$$

b =

$$
\begin{array}{cc}
 & \text{u1} \\
\text{x1} & 2.00000 \\
\text{x2} & 0 \\
\end{array}
$$

c =

$$
\begin{array}{ccc}
 & \text{x1} & \text{x2} \\
\text{y1} & 0 & 2.50000 \\
\end{array}
$$

d =

$$
\begin{array}{cc}
 & \text{u1} \\
\text{y1} & 0 \\
\end{array}
$$

The state-space matrices are then

$$
A = \begin{bmatrix} -6 & -2 \\ 4 & 0 \end{bmatrix}, \ B = \begin{bmatrix} 2 \\ 0 \end{bmatrix}, \ C = \begin{bmatrix} 0 & 2.5 \end{bmatrix}, \text{ and } D = \begin{bmatrix} 0 \end{bmatrix}
$$

The output is

$$
y(t) = C\underline{z} + D\underline{x} = 2.5\big(z_2(t)\big)
$$

Second, we can find the state-space model in another way. Let us define the states as

$$\frac{d}{dt} z_1(t) = \frac{d}{dt} y(t) = z_2(t)$$

$$\frac{d}{dt} z_2(t) = \frac{d^2}{dt} y(t) = 20x(t) - 8z_1(t) - 6z_2(t)$$

and the state and output matrix equations are then

$$\frac{d}{dt} \underline{z} = A\underline{z} + B\underline{x}$$

$$\underline{y} = C\underline{z} + D\underline{x}$$

with

$$A = \begin{bmatrix} 0 & 1 \\ -8 & -6 \end{bmatrix}, B = \begin{bmatrix} 0 \\ 20 \end{bmatrix}, \underline{z}(0) = \begin{bmatrix} 0 \\ 0 \end{bmatrix},$$

$$C = \begin{bmatrix} 1 & 0 \end{bmatrix}, \text{ and } D = \begin{bmatrix} 0 \end{bmatrix}$$

The output is

$$y(t) = C\underline{z} + D\underline{x} = z_1(t)$$

You can see that the two models are not the same by comparing the matrices that we obtained in both cases. But the plots will show that the two models produce the same response. The script to create the plots is EOCE6_13. The function that defines the system is the same as the function in the last example.

The plot is shown in Figure 6.32.

EOCE 6.14

Consider the single-input single-output system in state-space.

$$A = \begin{bmatrix} 0 & 1 \\ -8 & -6 \end{bmatrix}, B = \begin{bmatrix} 0 \\ 20 \end{bmatrix}, C = \begin{bmatrix} 1 & 0 \end{bmatrix}, \text{ and } D = \begin{bmatrix} 0 \end{bmatrix}$$

Derive the other representations.

SOLUTION

1. The transfer function representation

We will do that in two ways: one is analytical and the other is by using MATLAB.

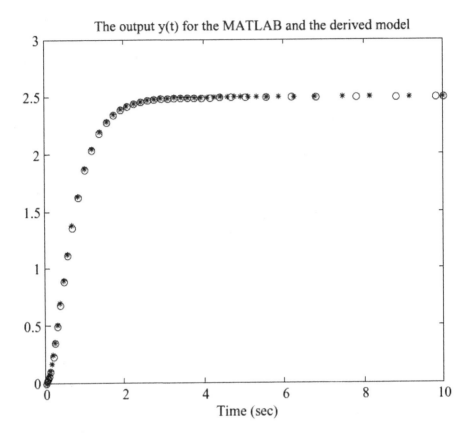

FIGURE 6.32   Plots for EOCE 6.13.

Analytically, consider the solutions of the state-space model in the Laplace domain

$$\underline{Z}(s) = (sI - A)^{-1} BX(s)$$

and

$$\underline{Y}(s) = C\underline{Z}(s) + D\underline{X}(s)$$

Substituting for $\underline{Z}$(s) in $\underline{Y}$(s) to get

$$\underline{Y}(s) = C(sI - A)^{-1} BX(s) + D\underline{X}(s)$$

and the transfer function is

$$\frac{Y(s)}{\underline{X}(s)} = C(sI - A)^{-1} B + D$$

We will substitute for the matrices and get

$$H(s) = \begin{bmatrix} 1 & 0 \end{bmatrix} \begin{bmatrix} \dfrac{s+6}{s^2+6s+8} & \dfrac{1}{s^2+6s+8} \\[2mm] \dfrac{-8}{s^2+6s+8} & \dfrac{s}{s^2+6s+8} \end{bmatrix} \begin{bmatrix} 0 \\ 20 \end{bmatrix}$$

And finally, the transfer function is

$$H(s) = \frac{20}{s^2+6s+8}$$

Using MATLAB, we write the MATLAB script

```
A= [0 1; -8 -6];
B = [0; 20];
C = [1 0];
D = 0;
[num, den] = ss2tf (A,B,C,D)
```

The output is

```
num =
```

$$0 \quad 0 \quad 20.0000$$

```
den =
```

$$1 \quad 6 \quad 8$$

Therefore, the transfer function is

$$H(s) = \frac{0s^2+0s+20}{1s^2+6s+8} = \frac{20}{s^2+6s+8}$$

2. The block diagram representation
   The system is second order and hence we need two integrators. The diagram is shown in Figure 6.33.

FIGURE 6.33 Block for EOCE 6.14.

3. The differential equation representation
   The state equation is

$$\frac{d}{dt}z = \begin{bmatrix} 0 & 1 \\ -8 & -6 \end{bmatrix}\begin{bmatrix} z_1 \\ z_2 \end{bmatrix} + \begin{bmatrix} 0 \\ 20 \end{bmatrix}x(t)$$

where

$$\frac{d}{dt}z_1(t) = 0z_1 + 1z_2(t) + 0x(t)$$

$$\frac{d}{dt}z_2(t) = -8z_1(t) - 6z_2(t) + 20x(t)$$

If we let $z_1(t) = y(t)$ and $z_2(t) = \frac{d}{dt}y(t)$ then

$$\frac{d}{dt}z_1(t) = \frac{d}{dt}y(t)$$

$$\frac{d}{dt}z_2(t) = \frac{d^2}{dt}y(t)$$

Now substitute in for the second derivative of y to get

$$\frac{d}{dt}z_2(t) = -8z_1(t) - 6z_2(t) + 20x(t)$$

or finally

$$\frac{d^2}{dt}y(t) = -8y(t) - 6\frac{d}{dt}y(t) + 20x(t)$$

6.9   END-OF-CHAPTER PROBLEMS

EOCP 6.1

Consider the following matrices.

$$A = \begin{bmatrix} 1 & 0 \\ 0 & 2 \end{bmatrix}, B = \begin{bmatrix} 3 & 0 \\ 0 & 4 \end{bmatrix}, C = \begin{bmatrix} 1 & 1 \\ 1 & 2 \end{bmatrix}, \text{ and } D = \begin{bmatrix} 1 & 1 \\ 0 & 2 \end{bmatrix}.$$

Find the following.

1. $(A + B)$, $(A - B)$, $(A - C)$, and $(A + C)$
2. AB, BA, AC, CA, ABC, and CBA
3. $\det(A)$, $\det(B)$, $\det(C)$, $\det(AB)$, and $\det(AC)$

4. A^{-1}, B^{-1}, C^{-1}, $(AB)^{-1}$, $(BA)^{-1}$, $A^{-1}B^{-1}$, and $(ABC)^{-1}$
5. The transpose: A', B', C', $(AB)'$, $(ABC)'$, $A'B'$, and $(AB)'C'$
6. Eigenvalues for A and B
7. Eigenvectors for A and B
8. e^{At} and e^{Bt}
9. Diagonalize C and D as P_1 and P_2, respectively
10. $P_1^{-1}CP_1$ and $P_2^{-1}DP_2$
11. Eigenvalues for C, $P_1^{-1}CP_1$, D, and $P_2^{-1}DP_2$

EOCP 6.2

Let

$$A = \begin{bmatrix} 11 & 2 \\ 0 & 1 \end{bmatrix}$$

and

$$B = \begin{bmatrix} 2 & 5 \\ 1 & 6 \end{bmatrix}.$$

Use the six methods described in this chapter to find

1. e^{At} and e^{Bt}
2. e^{ABt} and e^{BAt}

EOCP 6.3

Consider the following systems.

1. $\dfrac{d}{dt} y(t) + ay(t) = x(t)$

2. $\dfrac{d^2}{dt} y(t) + a \dfrac{d}{dt} y(t) + by(t) = x(t)$

3. $\dfrac{d^2}{dt} y(t) + a \dfrac{d}{dt} y(t) = x(t)$

4. $\dfrac{d^2}{dt} y(t) + by(t) = x(t)$

5. $\dfrac{d^3}{dt} y(t) + a \dfrac{d^2}{dt} y(t) + b \dfrac{d}{dt} y(t) + cy(t) = x(t)$

6. $\dfrac{d^3}{dt} y(t) + a \dfrac{d^2}{dt} y(t) + b \dfrac{d}{dt} y(t) = x(t)$

7. $\dfrac{d^3}{dt}y(t)+a\dfrac{d^2}{dt}y(t)=x(t)$

8. $\dfrac{d^3}{dt}y(t)+b\dfrac{d}{dt}y(t)=x(t)$

9. $\dfrac{d^3}{dt}y(t)+cy(t)=x(t)$

Write the state equations for each system but do not solve for the states.

Assume $y(0)=y_{00}$, $\dfrac{d}{dt}y(0)=y_{01}$, and $\dfrac{d^2}{dt}y(0)=y_{02}$. Under what conditions are the above systems stable?

EOCP 6.4

Consider the system shown in Figure 6.34.

Let $R_1 = R_2 = 1$ ohm, $L_1 = 0$ henry, $L_2 = 1$ henry, and $C = 1$ farad. Solve for the output $y(t)$ using state variables in the time domain and the Laplace domain if the input is an impulse of unity strength.

Is the system stable?

EOCP 6.5

Consider the system shown in Figure 6.35.

The rod to the left of the mass M can be modeled by a translational spring with spring constant k. The rod to the right of the mass M can be modeled by a translational damper of constant B. Let $M = 1$ kg, $k = 10$ N/m, and $B = 5$ N/m/sec. The equation describing the system is

$$\frac{d^2}{dt}y(t)=-10\frac{d}{dt}y(t)+x(t).$$

Solve for the output $y(t)$ using state variables in the time domain and the Laplace domain if the input is an impulse of unity strength.

Is the system stable?

FIGURE 6.34 System for EOCP 6.4.

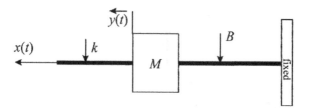

FIGURE 6.35 System for EOCP 6.5.

EOCP 6.6

Consider the system in Figure 6.36.
Let $R_1 = R_2 = 1$, and $C = 1$ farad. Solve for the output $y(t)$ using state variables in the time domain and the Laplace domain if the input is a unit step signal.
Is the system stable?

EOCP 6.7

Consider the system in Figure 6.37.
The rod above the mass M_1 can be modeled by a translational spring with spring constant k_1, in parallel with a translational damper of constant B_1. The rod above the mass M_2 can be modeled by a translational spring with spring constant k_2, in parallel with a translational damper of constant B_2.
Ignore the effect of gravity and let $k_1 = k_2 = 1$ N/m, $B_1 = B_2 = 2$ N/m/sec, $x(t) = \sin(t)$, and $M_1 = M_2 = 1$ kg.
The equation describing the system is

$$\frac{d^4}{dt}y(t)+6\frac{d^3}{dt}y(t)+7\frac{d^2}{dt}y(t)+4\frac{d}{dt}y(t)+y(t)=\frac{d^2}{dt}x(t)+4\frac{d}{dt}x(t)+2x(t).$$

Solve for the output $y(t)$ using state variables in the time domain.
Is the system stable?

FIGURE 6.36 System for EOCP 6.6.

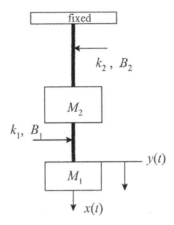

FIGURE 6.37   System for EOCP 6.7.

EOCP 6.8

Consider the system in Figure 6.38.

If $R_1 = R_2 = 1$ ohm, $C_1 = C_2 = 1$ farad, and $x(t)$ is the impulse function, solve for the output $y(t)$ using state variables in the time domain and the Laplace domain.

Is the system stable?

EOCP 6.9

Consider the system in Figure 6.39.

The differential equation describing the system is

$$\frac{d^3}{dt}\theta(t) = -(B/J)\frac{d}{dt}\theta(t) + (1/J)T(t).$$

The rod is attached to a fixed end and can be represented as a rotational damper only. The moment of inertia J represents the concentration of the mass M. A torque T is applied and it produces the angular rotation.

FIGURE 6.38   System for EOCP 6.8.

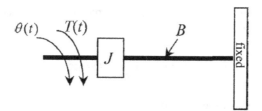

FIGURE 6.39   System for EOCP 6.9.

With $T(t) = 10$ N.m, solve for the output $\theta(t)$ using state variables in the time domain and the Laplace domain.

Is the system stable?

EOCP 6.10

Consider the system in Figure 6.40.

If $R = 1$ ohm, $C = 1$ farad, $L = 1$ henry, and $x(t) = 10\cos(2t)$, solve for the output $y(t)$ using state variables in the time domain and the Laplace domain.

Is the system stable?

EOCP 6.11

Consider the system in Figure 6.41.

Two rods with different masses will give rise to the two moments of inertia J_1 and J_2. If $J_1 = J_2 = J = 10$ kg.m$_2$, $k = 5$ N.m/rad, and the torque is applied for a very short

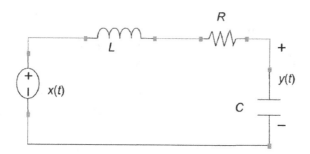

FIGURE 6.40   System for EOCP 6.10.

FIGURE 6.41   System for EOCP 6.11.

time that can be represented as an impulse, the relation between $\theta(t)$ and $T(t)$ can be described by the differential equation

$$J\frac{d^4}{dt}\theta(t)+2kJ\frac{d^2}{dt}\theta(t)=k\delta(t).$$

Find $\theta(t)$ using the real-time state solution.
    Is the system stable?

EOCP 6.12

Consider the system in Figure 6.42.
    If $R_1 = R_2 = 1$ ohm, $C = 1$ farad, $L = 1$ henry, and $x(t)$ is $u(t)$, solve for the output $y(t)$ using state variables in the time and the Laplace domains.
    Is the system stable?

EOCP 6.13

Consider the following rotational system.

$$2\frac{d^3}{dt}\theta(t)+\frac{d}{dt}\theta(t)+2\theta(t)=\delta(t)$$

Find $\theta(t)$ using the real-time state equation solution.
    Is the system stable?

EOCP 6.14

Consider the system shown in Figure 6.43.
    If $R_1 = R_2 = 1$ ohm, $C_1 = C_2 = 1$ farad, and $x(t)$ is $u(t)$, solve for the output $y(t)$ using state variables in the time domain and the Laplace domain.
    Is the system stable?

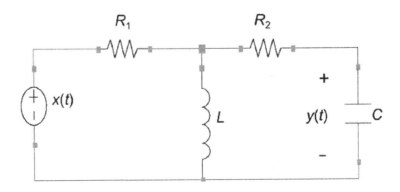

FIGURE 6.42   System for EOCP 6.12.

FIGURE 6.43 System for EOCP 6.14.

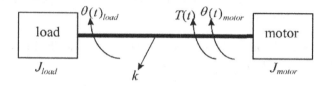

FIGURE 6.44 System for EOCP 6.15.

EOCP 6.15

Consider the system in Figure 6.44.

With $J_{load} = J_{motor} = J = 2$ kg.m², $k = 10$ N/m, and $T(t) = u(t)$, the dynamics that relate the angular rotation at the load side and the torque are given as

$$2\frac{d^5}{dt}\theta_{load} + 20\frac{d^2}{dt}\theta_{load} = 10u(t).$$

Find the angular rotation at the load using the state equations in real time.
Is the system stable?

EOCP 6.16

Consider the system in Figure 6.45.

FIGURE 6.45 System for EOCP 6.16.

The transfer function relating the input voltage $x(t)$ to the output angular rotation is

$$\frac{\theta(s)}{x(s)} = \frac{k_1/(RJ)}{s\big(s+(1/J)\big(D+(k_1k_2/R)\big)\big)}$$

where k_1 and k_2 are proportionality constants, and D and J are the viscosity and the inertia, respectively.

1. Find the differential equation representation.
2. Obtain the state-space representation.
3. Draw the block diagram.
4. Give values to the parameters shown in the transfer function. Obtain the impulse response representation.
5. Is the system stable?
6. Repeat by choosing other values for the parameters and obtain the impulse response until the system is stabilized.

EOCP 6.17

Consider the circuit in Figure 6.46.

1. Find the transfer function relating $X(s)$ to the current in the inductor flowing to the right.
2. Derive the differential equation representing the transfer function.
3. Obtain the state-space model.
4. Obtain the impulse response representation.
5. Use any representation to find $y(t)$ due to a step input of unity magnitude.
6. Draw the block diagram.

FIGURE 6.46 System for EOCP 6.17.

REFERENCES

Bequette, B.W. *Process Dynamics*, Englewood Cliffs, NJ: Prentice-Hall, 1998.

Brogan, W.L. *Modern Control Theory*, 3rd ed., Englewood Cliffs, NJ: Prentice-Hall, 1991.

Close, M. and Frederick, K. *Modeling and Analysis of Dynamic Systems*, 2nd ed., New York, NY: Wiley, 1995.

Cogdell, J.R. *Foundations of Electrical Engineering*, 2nd ed., Englewood Cliffs, NJ: Prentice-Hall, 1996.

Denbigh, P. *System Analysis and Signal Processing*, Reading, MA: Addison-Wesley, 1998.

Driels, M. *Linear Control System Engineering*, New York, NY: McGraw-Hill, 1996.

Golubitsky, M. and Dellnitz, M. *Linear Algebra and Differential Equations Using MATLAB*, Stamford, CT: Brooks/Cole, 1999.

Harman, T.L., Dabney, J., and Richert, N. *Advanced Engineering Mathematics with MATLAB*, Stamford, CT: Brooks/Cole, 2000.

Kuo, B.C. *Automatic Control System*, 7th ed., Englewood Cliffs, NJ: Prentice-Hall, 1995.

Lewis, P.H. and Yang, C. *Basic Control Systems Engineering*, Englewood Cliffs, NJ: Prentice-Hall, 1997.

Moore, H. *MATLAB for Engineers*, Upper Saddle River, NJ: Prentice Hall, 2007.

Nilson, W.J. and Riedel, S.A. *Electrical Circuits*, 6th ed., Englewood Cliffs, NJ: Prentice-Hall, 2000.

Nise, N.S. *Control Systems Engineering*, 2nd ed., Reading, MA: Addison-Wesley, 1995.

Ogata, K. *Modern Control Engineering*, 3rd ed., Englewood Cliffs, NJ: Prentice-Hall, 1997.

Ogata, K. *System Dynamics*, 3rd ed., Englewood Cliffs, NJ: Prentice-Hall, 1998.

Phillips, C.L. and Parr, J.M. *Signals, Systems, and Transforms*, 2nd ed., Englewood Cliffs, NJ: Prentice-Hall, 1999.

The MathWorks. *The Student Edition of MATLAB*, Englewood Cliffs, NJ: Prentice-Hall, 1997.

Woods, R.L. and Lawrence, K.L. *Modeling and Simulation of Dynamics Systems*, Englewood Cliffs, NJ: Prentice-Hall, 1997.

Wylie, R.C. and Barrett, C.L. *Advanced Engineering Mathematics*, 6th ed., New York, NY: McGraw-Hill, 1995.

Ziemer, R.E., Tranter, W.H., and Fannin, D.R. *Signals Systems Continuous and Discrete*, 4th ed., Englewood Cliffs, NJ: Prentice-Hall, 1998

Index

Note: Page numbers in *italics* indicate figures and **bold** indicate tables in the text.

Printed in the United States
by Baker & Taylor Publisher Services